Acoustic Justice

Acoustic Justice

Listening, Performativity, and the Work of Reorientation

Brandon LaBelle

BLOOMSBURY ACADEMIC
NEW YORK • LONDON • OXFORD • NEW DELHI • SYDNEY

BLOOMSBURY ACADEMIC
Bloomsbury Publishing Inc
1385 Broadway, New York, NY 10018, USA
50 Bedford Square, London, WC1B 3DP, UK

BLOOMSBURY, BLOOMSBURY ACADEMIC and the Diana logo are trademarks of Bloomsbury Publishing Plc

First published in the United States of America 2021

Cover design: Fliegende Teilchen
Cover image © Annette le Fort

Library of Congress Cataloging-in-Publication Data

Names: LaBelle, Brandon, author.
Title: Acoustic justice: listening, performativity, and the work of reorientation / Brandon LaBelle.
Description: New York City: Bloomsbury Academic, 2021. | Includes bibliographical references and index. | Summary: "Authored by leading sound studies scholar Brandon LaBelle, this book focuses on questions of acoustics as the basis for challenging normative structures"–Provided by publisher.
Identifiers: LCCN 2020034887 | ISBN 9781501368202 (hardback) | ISBN 9781501368226 (epub) | ISBN 9781501368233 (pdf)
Subjects: LCSH: Music–Social aspects. | Sound–Social aspects. | Sound art–History and criticism. | Justice (Philosophy) | Listening (Philosophy)
Classification: LCC ML3916 .L25 2021 | DDC 781.1/7–dc23
LC record available at https://lccn.loc.gov/2020034887

ISBN: HB: 978-1-5013-6820-2
PB: 978-1-5013-6821-9
ePDF: 978-1-5013-6823-3
eBook: 978-1-5013-6822-6

Typeset by Deanta Global Publishing Services, Chennai, India

Contents

Figures

Holding, Healing, Attending

Toward Collaborative Living

In the protests taking place in Lebanon starting in October 2019, one of the key aspects that has been continually highlighted is the opening up of sudden forms of togetherness—an overcoming or a breaking down of the deep sectarian divisions and separations that have marked Lebanese society and against which much of the uprising directs itself. Economic inequalities, the drastic privatization measures defining Beirut for example, and the grave lack of cultural infrastructures and public institutions outside of political parties and religious affinities have led to a general stagnation of civic life following the Civil War (1975–90). The uprising therefore directs not only itself against the current state of government but also in favor of nurturing an overall environment in which social equality, public culture, and togetherness are valued. This has been given expression through the capacity, within the dynamics of the uprising, to voice long-standing and often repressed feelings and sentiments.

> From early morning talk shows until the evening news, our physical landscape is completely colonised by [the political elite], their loyal pundits and the talk show hosts that welcome them with open arms on their programmes, often laughing and smiling. This has been the case for decades, not years. So much so that any average Lebanese can recognise the politicians' voices if the news is on in another room. But over the last week, that situation has changed drastically.[1]

The "open mic revolution," as Habib Battah outlines, is defiantly marked by a shift in acoustic dynamics: from centralized political media to the emergence of a plethora of voices, the "birth of civic society" in Lebanon brings forward demands that also work at shifting the dominant acoustic of division and regulation. Other more "independent" media channels have taken to the streets in their reporting, offering their microphones and cameras to protestors. As a consequence, the tonality of vocalized views and shared sentiments is shifting, drawing along in the sudden surge an awakening of social possibility and surprising cohesion. As one protestor stated: "We were at a place where there was a Lebanon for Muslims and one for Christians, but now Lebanon is simply Lebanon. The country is being born."[2]

The social and political transformations taking place gain traction by working at other forms of conversation, other forms of community building that have at their base a shift in acoustic and auditory norms. Following this example, struggles over economic injustices, and the fundamental operations of state structures to support its citizens turn upon a politics of the senses, where social recognition balances with the institutional arrangements enabling hope in the future. To "open the mic" is to already transform the dynamics of representational politics, upsetting the balance of power by giving way to new forms of social movement and articulation.

> The revolt has only gone on for a little over a week, but the pain and anger fueling it has been building for months and years. For myself and many others, it has been an incredible relief not to have to hear the news of politicians, the soap opera of their daily lives: their press conferences and trips abroad, their petty squabbles and self-victimisation, their patronising self-confidence, egoism and insincere promises. It has been a great relief to hear something different for a change, someone else's problems and some else's voice and to know that those voices out there, millions of them, do exist.[3]

The statements and sentiments capturing the situation in Lebanon point toward the questions and theoretical perspectives I work through in *Acoustic Justice*. I focus on how the capacity to "hear something different" is foundational to nurturing the ongoing work of civic life. While the dynamics of power and politics are shaped and reshaped by institutional systems and situations, they are also continually figured by way of daily and deeply persistent rhythms of care and consideration, indignation and desire. As such, power is found in the capacity to stay close to others, in the working through of what it means to be a body in an incredibly physical and dense world. I'm therefore keen to emphasize a notion of justice as what works at grounding power within the social spheres where common life finds its ongoing expression.

Acoustic Justice poses a view onto the ongoing birth of civic or common life by attending to the ways in which the power of people is figured and refigured through a relation to sound, audibility, attunement, interruption, and the arrangements we make to hear something different, as well as to *hear differently*. This, importantly, includes untethering hearing from the ableness of an ideal listener; rather, it is to understand "hearing differently" as a question of diverse sensing, orientation, and responsiveness, wielding what Jan Masschelein underscores as the "power of attention" as the power to engage with the present, to nurture the present as an arena of engagement and emergence.[4] In this regard, turning to sound as a medium and topic, as an acoustic question, I work at elaborating its force as one that also turns us toward each other in ever-demanding and ever-enriching ways. A turning that contributes to the power as well as to the paying of attention. The notion of "paying attention" is suggestive for understanding attention as an act, a gesture, and also an economy; it underscores attention as contributing to a balancing of power, of relational intensities and struggles, of ethical responsiveness and interpersonal engagement. In listening, as that moment of turning, or of being turned, one pays into the economy of attention, animating the social and common bonds to which one is indebted.

The concept of acoustic justice is posed here then as a framework for attending to the critical and creative ways in which individuals and communities work at power by seeking other configurations of the economy of attention—other orientations. Acoustics, as I set out to argue, is fundamentally about such questions of power, and the configurations one may work at in deploying or balancing such power. This is given expression, again, in the statements resounding on the streets in Beirut: "We don't have to listen to you minister, you have to listen to us now."[5]

Sonic Agency

Struggles over recognition, and the ways in which individuals and communities seek to transform existing systems of inequality or violence, may gain traction by way of sonic expressions: the act of raising the volume is often central to articulating dissent, mobilizing collective movements, and garnering the attention of governments and media alike. Such punctuations are equally expressed, at times, through silence, and the refusal to speak or to sound forth any type of demand or position: silence instead resounds, through a type of negativity, to upset the continuity of the usual and its tonalities. In this regard, sound is a fundamental material and means within social and political imaginaries and struggles, giving way to forms of sonic agency.

The outpouring of noise, or the insertion of a collective silence, is counterbalanced by recognizing the importance of listening and the experiences of being heard: how struggles over recognition, and the concomitant acts of listening these often entail, further detail the ways in which sound and listening participate in democratic procedures, judicial systems, and scenes of self-governance or shared resistances. Sonic agency may be articulated by way of volume, and the modulation of noise and silence, but it is equally the affective relations shaped by listening through which such volumes work at shifting the delimitations of power. Listening opens an important view onto understandings of agency by underscoring the political not as a drama of appearance and outspokenness alone, but rather, listening also reminds how recognition is effected by that sense of being heard, as well as the necessity to hear something different, where listening may also strike back, in terms of mobilizing people in their desire to reconfigure the flows of power, to demand a different form of life. While emphasis is often placed on the one who speaks, on the staging of the political voice, I'm interested equally to turn attention to the one who listens—the listener as an actant, and one that lends an important force.[6] For listening also draws us toward states of critical reflection, slowness, shared attunement, and capacities for understanding or care, all of which articulate other dimensions of power.

As a gesture, as an act, listening works at figuring scenes of support, enabling a process of collaboration that often extends beyond the apparent and the articulated. As such, it lends to the ongoing demands and challenges of responsiveness, to bring energy to acts of attention which, as Masschelein poses, open important "spaces of self-transformation."[7] Attention is a nurturing act, providing the basis for learning, for knowing, and for understanding. In this sense, it works at extending oneself: toward other knowledges, other viewpoints, other ways of living. I give my attention, and in doing so, I give myself over to what may emerge, to the emergence attention, in fact, makes possible. Attentiveness is therefore a critical and creative position, a stance one also takes in order to engage the collaborative work of living.

In listening, I would suggest, one intensifies a sense for the making of the present, flexing the borders of the spatial and temporal horizon—of this quotidian scene—with attentiveness: a tending which is necessarily a slow work. Listening, as an extremely key figuring of attention, though, is never stable or ideal; it is hard to maintain—it drifts, it falls, it lags behind, it is strained by what it hears—and that might be part of its essential contribution: it drags the dramatic theater of the political down. It requires time, as well as continual renewal and repair. The economy of attention, as that relational arena flexed by a range of forces, is animated, energized, and deflated by the ongoing ebb and flow of engagement and possibility; and as such, it indexes precisely the life of the body, its struggles and its pleasures. Listening may show us the very limit of ourselves, attuning one to the body's metabolism, along with the flows and rhythms defining our social bonds.

I stay close to listening, then, so as to highlight sonic agency as what may also turn us toward a limit of the political: toward the limit of a particular articulation of power, of recognition, and of support. In this sense, I focus on agency by way of listening, by way of the power of attention, as what figures scenes of transformation, as well as "civility."[8] As Étienne Balibar argues, civility may be mobilized as a means for challenging forms of violence by specifically supporting the "possibility of the political." The political, as what must work on behalf of a social community and just relations, for Balibar, is founded upon a culture of civility: the nurturing of dialogue and cooperation, of diplomacy, an art of negotiation or compromise which implies a sense for responsiveness, resistances, and accommodations. Civility, in this sense, must be worked at, cared, and fought for, in the maintenance of that political space, which, for Balibar, becomes essential for challenging the normativization of violence within our global times.

Following Balibar, civility draws out a sense for countering the ways in which violence becomes acceptable, even an option within the geopolitical environment: how the deployment of not only new techniques of extreme violence, but also the discourses and moral justifications, increasingly make violence acceptable in terms of what we come to expect of global affairs. In this regard, I'm interested to think toward the ways in which civility might be expressed, even materialized. In his work on technoscience and the posthuman, for example, Dimitris Papadopoulos identifies what he terms "generous infrastructures."[9] In tracking the movements of autonomous politics, movements found in a range of conflicted and creative sites, from migration and refugee communities to hacker and maker labs, Papadopoulos finds a culture of collaboration and cooperation. Importantly, such collaborative expressions span across national borders, linking up communities and languages to figure a compelling weave of agents and actants; this weave includes not only people, but equally material elements, the crafting and circulation of objects and tools, and a consideration of shared resources, all of which participate in movements (and imaginaries) of autonomy. In tracking these scenes or situations, Papadopoulos recognizes the "generous infrastructures" that come to emerge and facilitate the ongoing movements of such collaborative and cooperative dynamics. Generous infrastructures "allow for communities to maintain and defend the ontological conditions of their forms of life even when instituted infrastructures break down by failure or by intent."[10] Importantly, generous infrastructures draw connections between the political and matter, articulating a greater sense for the social and material currents that become the basis for building other worlds. Agile, connective, communal, technological—these infrastructures *are* generous by fueling the ongoing collective work of an autonomous politics.

The notion of generous infrastructures, as I'm interpreting it, may also be considered as *infrastructures of generosity*; they spirit a sense for the coexistence of forms of life, for human and nonhuman assemblages, for building alliances, and, as such, give precisely to supporting expressions conductive of biodiversity. By way of such generous infrastructures, one may start to detect strategies of civility aimed at fostering a culture of collaboration. Generosity, and generousness, might be appreciated as the connective thread holding people and things together, especially within scenes of autonomous and improvised lives. As AbdouMaliq Simone highlights in his essay "People as Infrastructure," within the seemingly marginalized or ruinous scenes of urban worlds infrastructures are regulated, and even materialized, less by state administrations and more by way of the "convertibility" and "conjunction" of people, materials, spaces, and energies.[11] "Such a conjunction of heterogeneous activities, modes of production, and institutional forms constitutes highly mobile and provisional possibilities for how people live and make things, how they use the urban environment and collaborate with one another."[12] Generosity, in this sense, is not so much a question of charity or gift giving; rather, it emerges within the patchwork of people and things, space and the improvised, yet regulated, flows and exchanges taking place. Generosity is therefore more socially functional and materially dynamic, feeding a general circuit of collaboration, and operating in a sort of "alchemical" way in keeping things moving.

I'm putting into play these different perspectives, from strategies of civility to generous infrastructures, in order to underscore my arguments around acoustic justice as being about the material processes and social practices that work at constructs of solidarity. In this regard, listening is positioned as the expression of the power of attention, one that may ground agency, and the enactments we work at, within the life of the body, as a life of imagination and interdependency—listening is what supports civility as a political perspective, as a culture of generosity. Acoustic justice is developed then so as to intensify an understanding as to the ways in which sound and listening impact onto the power and economies of attention and the articulations of political life, in the making of generous infrastructures. In this regard, it is wise to remember—

following the protestors in Beirut—that even while *not hearing* those "millions of voices out there," they do in fact exist.

Returning to questions of sonic agency, as that sense of agency driven by listening, and which works at balancing an economy of attention, at civility, I want to further consider in what ways "not hearing" is equally central to strategies of civility. Within the project of attention, as the drawing out of spaces of self-transformation and collaboration, listening is positioned not only in relation to the apparent and the legible, but equally to that of the missing: for what goes unsaid and unheard. The experience or capacity to "hear differently" is to bring recognition to the unheard precisely as being constitutive of the present; as being present in the negative, for if listening is to wield a power, it is to lead us toward the limits of the audible; it is to upset the "distribution of the sensible" which, as Jacques Rancière suggests, defines the shape of the political.[13] Listening, as part of the work of attention, educates how the present is deeply effected and touched by its own limits, by what exceeds or has gone missing from the arena of the sensible. In this regard, listening may attune one to hear the weight of the voice, its presence as well as its absence, those with *and* those without. It is to extend a relation to the voice, to the said, by recognizing what escapes the field of language, requiring that one relate to the verbal and the nonverbal, understanding these as always already linked.

Mojca Piškor's work on the rather little-known history of women's labor camps in former Yugoslavia in the 1950s poignantly underscores listening as a method for engaging precisely with histories marked by complex silences. As she suggests, "reading" into the archive of state violence is to hear the silences carried by those who cannot speak about past suffering, whose voices have been "confiscated" by the state.[14] For Piškor, listening becomes a form of reading, of critical and creative attentiveness, that assists in articulating often hidden or repressed experiences, and how one may work through the writing and rewriting of state narratives. While speech, and the outspokenness of truth, acts as a carrier of hope and the drive of emancipatory movements, it is necessary to remind of listening's role as what often fulfills the voice and the promise of the said. For listening, in this regard, specifically addresses the unsaid, attending to that which is withdrawn or repressed. In doing so, it further opens an extremely important sense for the not-yet—for what remains to be said. And as such, listening wields a deep intervention precisely within scenes of injustice by reconfiguring what we understand as presence and visibility, as the sensible.

Listening can dynamically shape, through a type of careful persistence, the temporal horizon of the political by allowing for a gap, a pause, a concentration, a reading: a diplomacy found within geopolitical debates, a criticality expressed in the gestures of research (to search *again*), as well as within the rhythms of daily practices, where listening immediately slows time, extending a space of concern, even of pleasure. Listening stretches us toward that which calls for attention, stretching in return a given frame of address.

This is not to overlook how listening can be utilized as part of forms of coercive power, where listening is invasive, penetrating certain scenes or arenas. Listening becomes a technique, a mode of capturing information, either in the service of emancipatory or exclusionary projects. There is a potential violence in listening, one that should not be forgotten. Listening is therefore put to work in different ways, figuring it within acts of justice and injustice alike.

It is my concern, here, to focus on listening as providing a deeply creative and critical path for enriching human relations, for nurturing and caring for the limits of the body and for struggles over recognition. By deepening or prolonging the silences surrounding speech, pausing along the way, figuring a space of concern, listening is a powerful gesture of attention, a path of attentiveness that works at refiguring the dynamics of recognition and acknowledgment. It is always a figuring of the present, a being in the present, a demand for engagement, yet one that carries the weight of the missing, of what is often said by being unsaid, and which radically assists in opening to a future to come—a listening that creatively hears more than what is spoken, directing us toward life in the making. As Fred Griffin outlines in his work on "creative listening" within psychoanalysis, one must bring an "active listening" to bear upon experiences of trauma and repression in order to move toward emotional growth.[15] For Griffin, developing a pronounced sense for listening to patients opens a greater appreciation not only for listening as a position of receiving but more importantly, for also allowing questions to be asked. By listening Griffin nurtures within his practice as an analyst the critical and creative capacity to engage more fully with the therapeutic process as a "collaborative" work: to enter the "space of self-transformation" as a process of shared growth. "I began to see this as an actively engaged, creative process that makes it possible to name dimensions of emotional life not fully articulated previously."[16]

Listening, as what brings attention to the said and the unsaid, slowing the time of speech with a greater sense for the missing and the previously inarticulated, to name those dimensions often hidden from view, assists in creating a "holding environment"—a supportive space of attention, a slowing to allow for dwelling upon the said: to draw out not only the meaningfulness of the spoken, but to also give room for its resonating silence. Listening, in this manner, "builds bridges" between the spoken and unspoken, the said and the unsaid; it draws into the arena of attention and understanding—not only through care and empathy but also through its collaborative capacity—its collaborative demand, nurturing an intensity of critical or associative thinking, for bridging knowledges, for thinking together, attending to the withdrawn *and* the not-yet spoken. Within such holding environments, the possibility to hear differently is given a chance.

This entails a relation to memory: within the holding environment, one also draws into play a sense for past experience. While history often marks out a genealogy of violence, and a subsequent politics as that which works at the stage of power, memory may figure a temporality of "anti-violence," a slowing down—a civility driven by repair and the necessity to keep in mind what has been lost, to mind that which is missing: the specters of lost lives. Memory, in this sense, reclaims time as being one not of progress made along a powerful line; rather, memory holds time, staggers progress, asking for a continual reworking of narrative—memory as a telling needing to be continually retold, reheard. (This is the fundamental difference between the memorial and the monument, attending to the ongoing question of memory or articulating history.) Listening, as a gesture of attention that holds, that figures a holding environment, also allows for remembering: to remember oneself as being human, as being full of emotion both accessible and inaccessible; it is to engage in the capacity to care, which is also effected by remembering what has been lost and what can be lost still. And further, it is to research the lines that constitute what we know of history, prying open the files and the records so as to figure a resonance, an amplification: a holding and a healing.

Listening, in this sense, might be to slow the culture of speed indicative of our contemporary productiveness—this endless event that drives continuous output. Listening, then, as a slowness may, as Isabelle Stengers has sought to remind, wield a potent resistance to the hurried pace of progress and the desire to make history. Her moving call for "slow science" is critically positioned so as to figure a time and space for affective experience and considered articulation—to allow the matters at hand to truly matter. Slowness for Stengers, as Nataša Petrešin-Bachelez suggests, is about "learning to listen to each other" in order to attend to the meanings and values that emerge outside or around objective categories.[17] What is at stake for Stengers is to combat the hurried production of knowledge and its impact onto the capacity to stay close to the creative process of inquiry. Slowness, and the importance of listening, thus gives way to allowing us to *not know*, to understand how not knowing is precisely the conditions needed for truly discovering the value of knowing.

Sonic agency—as that sense of agency shaped by the punctuated volumes of speech and noise, as well as the spirit of listening and remembering, concern and care—works at balancing the said with the unsaid, the knowing and the not knowing; it is posed here as a basis for the power of attention, as a strategy of civility which, to my hearing, is based on slowness: a hesitation, a strength of holding, a generosity so as to allow for answers and questions. Or rather, it may be to interrupt the political by demanding that it, in turn, listens to itself in order to remain valid to the ongoing struggles of being civil to each other.

On Acoustics: From Psychoacoustics to Social Acoustics

Acoustic Justice addresses questions of sonic agency by making a number of key moves in order to extend the work of listening as the basis for civility, or what Judith Butler terms an "egalitarian imaginary."[18] Central to *Acoustic Justice* is understanding listening as not always being bound to the audible, or to sound only (and distinct from hearing, which is more centered on the physiology of the ear). Rather, listening, as I've articulated, is posed as a fuller expression of attention, as a multimodal, embodied power that enables a range of conversations and contacts, intelligence and imagination. Listening, in this regard, integrates hearing within its activities and capacities; it draws from hearing, as a physiological operation, and as a general sensing, while focusing and extending its reach. Such an understanding of listening is suggestive for approaching acoustics. On one hand, acoustics is understood as the physical conditions, the architectures and spatial arrangements, that facilitate and shape the reflections and reverberations of sound: acoustics as a question of the physics of sound, the materiality of space, and the physiology of hearing. Following this perspective, acoustics dramatically contributes to a sense of orientation as well as belonging, lending to how we navigate through spaces and environments in capturing a sense of place: how we synchronize, attune, and align with others. From such a base, acoustics is underscored as effecting experiences of hearing and the expressions of sociality, on the relationships we may form and in which listening becomes more operative. This finally leads to considering acoustics as having an impact onto a politics of recognition and location, and articulations of forms of life: acoustics as a politics through which struggles over recognition and rights, belonging and access are waged.

Rather than a strict concentration on sound only, and the punctuations it enables, I draw out acoustics in order to focus on the conditions that surround and that effect bodily and social life, on the affective and political dynamics shaping our situatedness. This entails a concern for technology, and its place within acoustic experiences and methods—not only in terms of the construction and distribution of recorded sound but also in terms of how technologies impact subjectivity and approaches to ability, for instance within Deaf culture. In this regard, focusing on acoustics enables a range of critical and creative inquiries, which have at their center a concern for orientation and how we may work at reorientation, the ways in which one navigates and negotiates the systems and discourses that impact onto defining one's place in the world. While sonic agency is expressed by way of the articulated and punctuated sounds one may make, acoustic justice is figured by considering the arrangements and configurations that allow for different types of orientation, from social and political to bodily or sexual. If listening is about the power of attention, about an economy of attention and the potentiality of emergence, then acoustics is what we do to facilitate or modulate that economy.

Acoustics is thus posed as a question of material and social relations, as well as the practices and gestures that work at forms of orientation and reorientation: the architectural and spatial arrangements and social scenes, the vocal articulations or interruptions, the technological systems or cultural expressions we make, in order to capture forms of life. This moves us from the psychoacoustic, and the experiences and imaginaries of hearing, to the social acoustic, and the dynamics of life with others, and equally from the electroacoustic, as the mediations of distributed sound, to the bioacoustic, as acoustic ecologies of human and nonhuman life, and that informs conceptualizations of *life by way of hearing*. From interrupting the dominant tonalities of places, shifting the rhythmical organization of socialities, or sounding out other forms of speech and affection, acoustics is mobilized to give traction to a diversity of voices and bodies, desires and endeavors—to figure just relations. Acoustics, in other words, is that which we do in order to reorient the articulations of sociality and subsequent forms of life. Acoustics thus emerges as architectures, as material and social arrangements, as personal and political acts, as well as through listening and the wielding of the power of attention, sentiment, a point of view, even a discourse: as what works at shaping and reshaping relationships.

Throughout *Acoustic Justice* I work at elaborating this expanded understanding of acoustics. This leads to both addressing experiences of sound and listening, as well as challenging us to use acoustics as a bridge to other experiences, topics, and questions, from issues of language and voice to sensory politics and bioethics. In this regard, I mobilize a critical poetic approach, maybe even a slow approach, in so far as terms such as "acoustics" and "sound," "rhythm" or "vibration," work both literally and figuratively. This is intentional, as my interest is to specifically unfold acoustics, and acoustic justice, as a generative framework, lingering over the particular movements of sound and the ways in which listening moves us, holding, nurturing, exciting us, as well as being figured within a range of political struggles. In short, I'm looking to put acoustics on the move so as to equip our listening sensibility with creative and critical tools—with a deeply mobile imaginary. For what is at stake, as I see and hear it, is to unpack the embedded potentialities and speculative possibilities found in an acoustic model, as that which enables diverse forms of orientation to take place, and, in taking place, may also enrich the meanings taking place might come to engender: how taking place, by way of acoustics, is equally a nurturing of what place can hold, from the place of the book, and the spaces of thought, to the places where forms of life express themselves, root themselves, and, further, to those places still *under construction*.

In working through a framework of acoustics, from the psychoacoustic to the social acoustic to the bioacoustic, I draw from a range of references from Sound studies, as well as social and political theory, that allow for thinking acoustics in an expanded sense. These include the legacy of soundscape studies and acoustic ecology, the research center Cresson and their related theories of sonic effect, and the work of Barry Blesser and Linda-Ruth Salter, whose work on aural architecture provides a rich analysis and compendium of acoustic-related topics and understanding. While these three references are important in my approach, I do not engage in unpacking the many aspects and questions they pose; instead, I recognize their research as a general ground, one that I draw upon for support while also extending away from. Blesser and Salter's concept of "aural architecture," for example, sets out to integrate the personal and affective experiences of listening as part of the more objective analysis of sound often found in understandings of "acoustic architecture."[19] While Blesser and Salter shift from a science of acoustics to that of aurality, and the cultural interpretations of the heard, I'm interested to further work at acoustics as a framework for questioning its place within struggles over recognition and belonging. This concern is also partly found in acoustic ecology, which theorizes the soundscape as the state of environmental sound and how it signals a range of social, political, ecological, and cultural dynamics. Research into the specificity of a soundscape, for example, may lead to recognizing inequalities or imbalances of power, linking sound and hearing to questions of community.[20] I equally follow this perspective, yet also work at complicating the soundscape as never always being about sound and the audible; rather, the soundscape is seen to implicate a range of other matters and materials, systems and senses, and therefore calls for greater attention to the inaudible and the non-hearing. For acoustics, in emphasizing the physics of sonic phenomena and behavior, opens the sound spectrum as being addressed to the hearing world, but also where vibration, pressure, and frequencies above or below the range of human hearing are equally active. In doing so, a science of acoustics also allows for extending conceptualizations and understandings as to what constitutes the body and the socialities by which subjectivity is figured. As such, I search for ways to activate acoustics in order to consider a greater sphere of experience, from the movements and sonorities of the nonhuman to the force of matter, the tactile experiences, and felt encounters that often exceed the hearable, the knowable, and the recognizable. Aural architecture and acoustic ecologies are thus frameworks for posing a critical tussle with a politics of recognition: for unpacking soundscapes as implicated within balances of power.

The work at Cresson, a research center at the National School of Architecture in Grenoble, additionally aims at elaborating a view onto the experiences of sound, and through the concept of "sonic effects" offers an important interdisciplinary perspective.[21] From such research, acoustics is mobilized in suggestive ways, spanning architecture and sociology, music and science, thereby situating sound and listening as a foundation for knowledge. By way of acoustics, I also argue for concentrating on the impact of sonic experience and subsequent forms of sensing and knowing. In this regard, I work at questioning how experiences of listening are affected by acoustic decisions and norms, and which greatly influence the shaping of knowledge. For my argument, acoustics is positioned to include questions of social and physical orientation, and is thereby implicated in a system of cultural norms and values that act to frame our experiences, our bodily and social figuring. As such, I'm posing acoustics as what includes both the sensed and the systemic, physical materiality and institutional orders. In this way, the concept of acoustic justice is proposed as a framework that may engage acoustic ecology, and the theories of sonic effect and aural architecture, mobilizing these knowledges toward questions of justice and struggles over recognition.

Returning to Jacques Rancière, and his concept of the distribution of the sensible, I'm interested to underscore acoustics as a political question. As such, I propose to consider acoustics as the "distribution of the heard," as what impacts onto the movements and shaping of sound and subsequent experiences or capacities of listening. As the distribution of the heard, acoustics gives articulation to economies of attention and the conflicts over acoustic norms and values that effect physical and social orientation. For instance, the ways in which one's attention is shaped in particular ways, directed or oriented according to certain values or social habits, and through related systems of technology or political economies, reveal acoustics as impactful onto what we hear and, also, how we hear it: how we come to access or participate within certain acoustic arenas. And further, how listening is always already a question of being a subject situated within particular contexts and communities. In this regard, soundscapes, aural architecture, and sonic effects are already a question of the distribution of the heard, as what defines the movements and articulations of sound and listening, as well as the capacity for orienting and expressing forms of life. To hear differently then is to work at reshaping acoustic power.

Silvia Federici highlights our experience of being a living and sensing body as something that has been continually shaped by capitalism. "Indeed, one of capitalism's main social tasks from its beginning to the present has been the transformation of our energies and corporeal powers into labor powers."[22] Transforming the body as a sensual power, and labor force, by way of a range of technical, mechanical, and scientific projects throughout the history of capitalism, from Taylorism and mechanization to computational and genetic models that figure the body as so many atomized parts requiring continual regulation and administration, for Federici indexes the constant need to reclaim the body. "Our struggle then must begin with the reappropriation of our body, the revaluation and rediscovery of its capacity for resistance, and expansion and celebration of its powers, individual and collective."[23] To reclaim and celebrate the body is to honor much of its inherent power of sensing and knowing, moving and making—a deeply energetic, creative intelligence that, following Federici, is central in the struggles against exploitation and the drive for a more egalitarian world.

Acoustic justice is about expanding upon listening as an extremely dynamic expression of bodily power, as a sensual and deeply transformative capacity by which to express individual and collective understanding and collaboration. Listening is a broader capacity to attune and attend that supports processes of reflection and empathy, compassion and care, for oneself and for others, which may assist in contending with the systems around us. As Federici poses in her argument on the need to reappropriate the body: "Our bodies have reasons that we need to learn, rediscover, reinvent. We need to listen to their language as the path to our health and healing, as we need to listen to the language and rhythms of the natural world as the path to the health and healing of the earth."[24] Listening is captured here as the means by which to learn the languages of the body, to attune to its inherent rhythms as paths of power and knowing, as well as healing. Within the conflicts over forms of life, and the biodiversity expressive of a common world, listening is wielded as a capacity to contend with genealogies of capture and enclosure by explicitly forging a path—an acoustic frame by which to cultivate more considered approaches for being on the planet.

On Justice: Recognition, Wildness

Positioning acoustics, and the experiences of sound and listening, as a question of justice requires some reflection on understandings of the just. Questions of justice draw into play a range of perspectives, including law and legal practice, philosophical traditions on moral responsibility and equality, theories on the ethical or good life, along with the more concrete social movements that engage the issue of rights. All of these contribute to understandings of justice, and, as such, reveal justice as foundational to ideals of human society, the moral and ethical ground by which law and legislation is often understood or expected to work, as well as what we often draw upon in working against inequalities or scenes of violence and harm. Justice therefore acts as a guiding imaginary and principle—a set of principles or desires around which law as well as social life and experience is positioned, in which equality, fairness, and the understanding of mutual freedom are continually addressed. In this sense, as George P. Fletcher suggests, Western democratic societies, for example, are founded upon an ongoing struggle or public debate over how best to construct a just society.[25]

Justice, in this sense, emerges as an operation and a politics, but also, a collective imaginary, a discursive and emotional framework around which struggles over equality, freedom, opportunity, and recognition play out across society. It is my interest to reflect upon the ways in which understandings and modes of sounding and listening participate in such struggles, such imaginaries, and further, how a theory of acoustics may contribute to figuring justice as an articulation of ethical and collaborative life, as what we grapple with in terms of understanding responsibility and responsiveness to others.

In considering justice, I take particular guidance from the work of Dimitris Papadopoulos, and, in particular, what he terms "immediate justice."[26] For Papadopoulos, justice is to be found not necessarily by way of an appeal to forms of retribution, by way of criminal courts and a politics that takes place in a future to come. Rather, Papadopoulos considers justice, following Walter Benjamin's thoughts in "Critique of Violence," as something that happens in the moment: that takes place now, within and by way of the conditions people work through or around. This form of justice is located "independent of law," in so far as it aims at exiting the cycle of violence articulated in the dialectic of "law-preserving" and "law-making," that is, between a state power of administration and the oppositional movements that enact violence only to instate a new form of law. This is understood by considering the inherent violence enacted by law itself, by the act of laying down the rules; such an act entails a level of violence and obedience, impacting onto the ways in which life may take place, as manners of living. This further presumes, by way of an administrative operation, acts (or ways of life) that break the law, and, in doing so, violate the rules of a social and political order. Working to carve an opening beyond this dialectic, the cycle of law-preserving and law-making, immediate justice, as Papadopoulos suggests, is realized in the "space of the ordinary"[27]—those spaces grounded in "the radical making of alternative forms of life and everyday materialities that exist outside the law and outside the eternal cycle of state violence and oppositional destructive violence."[28]

Immediate justice, though, is not aimed at forms of vigilantism, of taking the law into one's own hands; rather, immediate justice works at grounding justice, not as an abstraction, as a rule book, or as the articulation of judgment, but, rather, as a milieu, a landscape—what Andreas Philippopoulos-Mihalopoulos also terms as "lawscape."[29] The lawscape is posed as the intersection of law and space, a spatial turn within law that captures the sense in which law is distributed spatially, and how space is always already a question of law: the codes by which space becomes inhabitable, divided, contested, and shared. If law is the *laying down* of rules, it inscribes itself onto a social body by delineating it, demarcating it, marking the ground as the very territory in which law establishes itself (as "the law of the land"). Importantly, for Philippopoulos-Mihalopoulos the lawscape is a geography open to contingency, to the ways in which "visibilization and invisibilization" as he calls it—the ontological shape of the lawscape—is continually at play, worked and reworked, articulated or contested in everyday life. In the spatialization of law, law becomes a ground, a territory, and articulates more overtly its inherent relation to bodies and manners of living, architecture and the city, from national territories to daily geographies. As such, the lawscape grants a "maneuvering space" to how one may move through law: interrupting its codes, navigating its delineations, reorienting its ordering, modulating its visibilization and invisibilization. This finds an echo in Giorgio Agamben and his work on "forms-of-life." As he suggests, "It is not justice or beauty that moves us but the mode that each one has of being just or beautiful."[30]

Returning to the notion of immediate justice, the spaces of the ordinary that Papadopoulos highlights express that sense for the ways in which law and justice are lived, grounded or brought home, how they frame us, and, as such, are negotiated, given ground or restricted, challenged, and that does not necessarily entail the textual codes of law or an act of final judgment. Here, I may return to consider Balibar's "strategies of civility" as what may support the ordinary working through of relations of justice; strategies of civility that, within the here and now, on the grounds of common life, may materialize in diverse ways, and which give shape to expressions that neither keep nor break the law—but rather, walk through the lawscapes of the everyday.

Immediate justice, for Papadopoulos, is particularly crucial for reorienting justice from out of the liberal constructs of humanism that are necessarily built upon legacies of coloniality. Here, we might think of the ways in which "freedom" and "equality" are, mostly, articulated upon conditions and systems of inequality.[31] A decoloniality, as a project, would then take aim precisely at the "universalisms" of modernity that maintain radical asymmetries between human lives. "Decolonial movements . . . challenge the universalism of humans and the perpetuating injustices that were set up through colonial modernity and still split humans among themselves."[32]

With a focus on immediate justice, Papadopoulos asks for a deeper, more unwieldy tussle with the ongoing injustices riddling the planet. To engage more fully in the project of decoloniality is to "restore justice" within the immediate ecologies humans and nonhumans inhabit together: daily ecologies that span across forms of life and matters of coexistence, and that are necessarily complex, uneven, and tensed. Immediate justice is therefore a "posthuman justice, the co-construction of life with other species and objects, the simultaneous emergence of ethos and ontology."[33] Immediate justice is a "thick justice," as Papadopoulos suggests; it is expressed by way of the thick arrangements and entanglements of planetary life, of social and material assemblages, human and nonhuman intersections, and where articulations of equality must be elaborated, put into question.

Following these ideas of immediate or thick justice, I strive at articulating acoustic justice as a base for daily practices and gestures that work at just relations, and that take guidance from how we come to hear each other, or how listening may enable a greater sense for acoustics as a craft: the cultivation of sensitivity for what's at stake in being heard. To do so, I also consider the radically diverse range of "voices" that come to populate the field of hearing—voices of nonhuman species and the felt encounters with things, technologies, and networks, all of which force human hearing into acoustic ecologies that are necessarily excessive, *thick*, moving by way of frequencies, vibrations, tonalities, and rhythms radically within and beyond human sociality. In this sense, I suggest that acoustics acts as a dynamic means for nurturing interpersonal sharing, as well as contact with the nonhuman, where listening opens toward a bioacoustic, or even "geoacoustic" view. The sound spectrum does not distinguish between human and nonhuman, material and immaterial, voice and noise, technology and bodies, but, rather, is constitutive of a flux of energies, movements, systems, and languages, figuring listening as radically open—listening as prone to radical openness, looping by way of the skin, and the deep touch of the vibrational and ambient contours of place, of feeling place, and toward the voices and articulations that impart so much information and emotion, desire and vitality, including the animate and inanimate entities and machines that make noise, that distribute noise, and, further, to the energies (and "messages") of the transmitted, the networks, and the potentialities of listening across scales, from the micro to the macro. From within such a complex scene of material force, which is equally a scene of cultural and political power, listening is poised, hinged, troubled—forced open, at work.

As part of this view, I pursue acoustic justice as a vocabulary for addressing the fundamental need for recognition, as a social mechanism. Recognition is positioned as what is partly at stake in the acoustic practices and arrangements, attunements and interruptions we make, and that facilitate or obstruct the capacity for giving and receiving attention: for finding support within the density of the world. Whether within our most intimate relationships, or as part of one's sense for participating within a greater community, attention and acknowledgment of others greatly impact onto struggles over recognition. Yet, in following acoustic justice, I also ask: What becomes of social recognition when considering a greater planetary and posthuman view? What does recognition mean within acoustic ecologies of human and nonhuman life? If listening is prone to radical openness, to contending with explicitly diverse forms of life, might it challenge understandings of recognition, specifically enabling a more fundamental engagement with the unrecognizable? And in what sense can acoustics, and acoustic justice, relate to issues of Deafness, where "having a voice" is no longer captured by way of sound?

In his work on recognition, Axel Honneth provides an important theory for understanding expressions of equality and solidarity, and which, along with Papadopoulos, provides a general reference for my own approach to thinking about justice. By way of a careful reading of Hegel, Honneth arrives at three forms of recognition. First, recognition emerges within the family, and in particular, through the relation between parent and child, giving way to a recognition of need and the articulation of love. Importantly, these fundamental experiences are carried forward, into mature forms of intimacy and caring. Second, within a framework of law, persons are recognized as the bearer of rights, as subjects respected through having access to making claims. As a citizen, I am recognized as a subject. Finally, recognition is found within the sphere of social life, where one is recognized as having worth within a community of values—as a person seen as making a contribution, as having something unique to offer. These three forms of recognition, for Honneth, give way to articulations of love, equality, and solidarity: from the affectional attachments and bonds of love, to the recognition of each other as subjects with equal rights, and, finally, to the solidarity gained by recognizing the social value of each person, and what each may bring to the community. Recognition therefore comes to articulate a "moral respect" through which I see myself through others, and others through myself. As such, feelings and experiences of disrespect—as that sense of recognition being withheld, which underpin experiences and systems of exclusion—become the ground from which social and political struggles emerge: to struggle for recognition and, therefore, for justice or just relations. As Honneth argues: "The forms of recognition associated with love, rights, and solidarity provide the intersubjective protection that safeguards the conditions for external and internal freedom, upon which the process of articulating and realizing individual life-goals without coercion depends."[34]

What's important in Honneth's model of recognition is how recognition is a binding social mechanism, underscoring "the role of intersubjectivity within public life."[35] Through such intersubjectivity, in which one obtains a sense of self-respect and self-esteem through the recognition granted by others, public life may flourish, specifically tending toward ethical responsiveness. Yet, importantly, recognition is highlighted as being based upon conflict—recognition, as Honneth underscores, is a struggle and, accordingly, it continually passes through stages of disappointment, interruption, and loss. Recognition is therefore a mechanism shaping the dynamics of attention and acknowledgment, and the flows and rhythms of being-in-common as what we work at, and as what works us.

Extending from Honneth's theories, the concept of acoustic justice is positioned through a general framework of recognition, which leads to questioning particular situations and forms of social struggle and cultural expression. While Honneth maps an important overview of the major forms and processes of recognition, I'm also concerned to underscore recognition as an emergent process that finds expression in the small acts of everyday life; recognition as found in moments of encounter, which continually rides upon the shifting dynamics of relationships. This understanding seems extremely important, to remind of the deeply informal currents underpinning love, equality, and solidarity. Such a view runs throughout my thinking on acoustics and acoustic justice, moving across both formal and informal modes of recognition. Recalling the demonstrations taking place on the streets in Beirut, acoustics emerges not only as a property of space, but more as gestures and practices that support struggles over recognition, that affect our drive for love, equality, and solidarity—the desire for more pronounced forms of civic life and social togetherness—by specifically modulating the tonalities of places, the economies of attention, and the norms by which one may gain a voice, obtaining access to social and political communities. In short, movements that also ground justice, searching for routes beyond an ongoing cycle of violence and injustice, and that may also spirit a new sense for living together.

Importantly, I also work at complicating understandings of recognition by considering acoustics as supportive of ecologies of human and nonhuman life, of animate and inanimate matter; as thick posthuman ecologies in which recognition is upset or challenged by way of the nonhuman and the nonliving—which, in the case of sound and listening, extend across ultra- and infra-frequencies and registers, from the bioacoustic to the electroacoustic, from bodies to machines. Or, the ways in which things resound and continually punctuate the soundscape, as well as our psychoacoustic life—humming as the general noise of the world. Such a view leads to question the ways in which hearing takes place within the immediate ecologies humans and nonhumans inhabit, and which include the hearing of other life, other things, as well as recognizing one's own limited capacity to be heard. This also entails a consideration of the ways in which sound is never always heard, but rather, is articulated equally through feeling, movement, and visualization, which become highly operative within Deaf culture.

In what ways do economies of attention, and the nurturing of love and solidarity, include relations with things, with nonhuman species, and the matters necessarily co-constitutive of those economies? In this regard, I follow Honneth's work on recognition by also questioning what happens when confronting the greater arena of audibility and sonic experience, the acoustic ecologies and aural architectures of posthuman experience. By way of an acoustic justice framework, I'm curious to follow how contacts and affinities that extend or upset human sociality articulate other forms of recognition, and other ways of understanding love, equality, and solidarity. For instance, what of the love passing between species, between humans and animals, or the sense of equality found in recognizing oneself as part of the nonhuman, as being vulnerable to and dependent upon so many organic entities, such as food, and the solidarity we may develop with things, with musical instruments, our computers, with soil, and with time itself? Do we not establish deeply powerful and rich affinities with such things? And do not these things also take from me, drawing me into their worlds? In this sense, I'm also interested in what ways acoustic justice, as the figuring of that arena of audibility, and listening's attentive reach, relates to what Cormac Cullinan terms "wild law"—the making of a system of human governance directed by a planetary perspective.[36] Wild law works at extending our human-centered systems beyond a framework of social justice and established modes of recognition toward Earth justice. The notion of "the wild" for Cullinan points toward the deeply creative and self-regulating systems of nature, and which may also reorient our understandings of "civilization" to which law is often based. Wild law is positioned so as to foster greater acknowledgment and attunement—even a wild embrace!—with Earth systems so as to move human governance into participating in the flourishing of biodiversity rather than instigating its collapse.

The Right to Listen

While sound and hearing, voice and listening are key to more formal practices of justice, for example within traditions of the legal trial that are dramatically centered around the articulation and hearing of spoken arguments, I bring focus more onto other situations and scenes. Yet, in doing so, I also borrow concepts from law and, in particular, the construct of rights central to struggles for justice. For instance, the principle of reply, or the right to be heard (*Audi alteram partem*, "to listen to the other side") found in legal systems and judicial proceedings may be deployed in considering acoustic justice as a condition that not only ensures one has a voice but also necessitates a state of listening. Extending from such principles and acoustic dimensions, I equally speculate on what a "right to listen" might suggest in terms of listening's role within justice work. In this sense, I borrow from the discourse of law and rights, and the operations of the fair trial, as a formal scene of voice and hearing, for these can be appreciated as informing a greater collective imaginary for how speaking out and hearing each other are fundamental to forging a just society, a just system.

In his essay "The Soundscape of Justice," the scholar James Parker maps out what he calls "acoustic jurisprudence." Through this concept, Parker argues for a greater consideration of the ways in which sound and law intersect, which is expressed through the physical architectures of legal settings, for example, as well as by way of the voice and its place within the history of law—the voice being key to a legal decree. Acoustic jurisprudence, as Parker suggests, is "concerned with how law is lived, both in sound and by virtue of it."[37]

Parker elaborates acoustic jurisprudence through a detailed account of the trial of the singer-songwriter Simon Bikindi held at the International Criminal Tribunal for Rwanda (2006–8). By examining not only the question of voice, but in particular, the "judicial soundscape" of the courtroom, including the mechanical systems facilitating speech as well as translation and interpretation, the use of headphones and microphones, along with forms of soundproofing, for example, Parker maps an important view onto the significance sound plays in legal settings. Through questioning the acoustic dynamics of the trial, he underscores the ways in which the judicial system is shaped by an acoustic performativity, where the structures that make formal articulations of justice possible are bound to a complex sonic configuration and apparatus. Importantly, this comes to include what Parker terms "law's sonic imagination": the engrained habits and values through which law resounds, is made audible. The staging of arguments, and the presentation and analysis of evidence, for example, can be underscored as highly situated, where one must work through the given constructs of legal and social systems and how such systems imagine fairness taking place: how the "habitus" of law delimits the spoken and the heard, directing listening in particular ways.

A fair trial as a defining feature of the courtroom, and the procedures of a hearing, brings forward an extremely poignant arena for querying the ways in which understandings of justice, and the practices that define access to legal rights, are partly if not significantly bound to a sonic or acoustic imaginary: that formal enactments of justice, as with other arenas of democratic society, presume certain methods or modes for the presentation of arguments as well as the right to be heard or to respond. As such, I am moved to stress acoustic justice as something that also migrates in and out of the legal system—taking place *on the side of law*—where experiences or processes of address and redress, the quotidian trials, for instance, found within spaces of education, such as the classroom or the lecture hall; spaces of resistance and autonomous politics, such as the self-organized center; scenes of medical attention, such as the clinic or the hospital, all come to draw out the question of acoustics as the shaping of audibility, the sensing of being in or out of place, of being part of a system that works fairly or not. Here, I may recall the notion of the lawscape posed by Philippopoulos-Mihalopoulos, casting it as an acoustic dimension, a reverberant terrain—lawscape equally as a soundscape by which one negotiates or generally inhabits "the rules," and by inhabiting, navigates their borders. How one's voice, for example, is registered, given value, or positioned within the distribution of the heard comes to reflect the complex and highly charged acoustic dimension always already at play. And yet, acoustic justice also works by way of the materiality of sound, from the bioacoustic to the electroacoustic, which greatly extends beyond that of human sociality, and therefore, human legality. In the thick, posthuman justice model I'm mobilizing, the "rules" of the soundscape are shaped by way of an extensively dirty, or wild ontology. That is to say, inhabiting the soundscape is to always already *be interrupted* by an excessive, dynamic, energetic, and diverse geoacoustic community.

In this respect, I'm concerned to question in what way "being heard" takes place, is given place, and works at reorienting a relation to place. Returning to Silvia Federici's call "to listen to the body" as a path toward resistance and healing, I might ask: How does such listening take place—*how do I listen to my body*? If I understand "my body" as an acoustic chamber, as something to be heard, how does it give way to such listening, accommodating or resisting it? And in what ways is such listening challenged or undermined by my understanding of listening, by my own cultural background or the social systems that also shape or direct my listening? Or by way of technology, which has always participated in defining listening's reach and abilities? Further, what might such listening generate or engender—how to carry this listening into the world? In listening to my body, wouldn't my body reveal itself as being constituted by both human and nonhuman vitality, an ensemble of materiality, calcium, bacteria, blood? And if so, does not reclaiming the body entail hearing differently, by way of such posthuman acoustics, unpacking the body as more or less than human?

Struggles over recognition often find traction by intervening upon the conditions that define hearing and being heard, that regulate one's attention and orientation with respect to oneself and others—that, in many ways, situate the fair trial on the street, in the home, in the spaces and temporalities between friends and strangers. It is these interventions that are of concern throughout *Acoustic Justice*, and which leads me to understand acoustics, or more specifically, acoustic justice, as those things we do in order to make listening to the body and each other possible, and that we may carry further, to underscore the importance of hearing a diversity of views across the public sphere.[38] In this regard, I may speak of "acoustic rights" or principles as not only related to the right to free speech, or the right of reply, as human or civil rights, but also related to what may encapsulate a broader sense of acoustic justice, to include setting the conditions not only for a freedom of expression but also a freedom of listening. This may be articulated along a number of lines, such as the right to listen to each other, as in the sharing and circulation of life stories, and which, as I've suggested, not only attends to the said and the articulated but also draws out a critical and collaborative space of concern for that which is missing, enabling greater sensitivity for the unsaid or the unspeakable (as well as the unable). Such an approach may also support moving from nurturing human relationships, and elaborating a diverse public discourse to acknowledging ecologies of human and nonhuman life in a caring or wild manner. A concentration on acoustic rights may entail providing education, for example, on listening as a practice, a skill, as the capacity for enriching empathy, understanding, affection, as well as critical inquiry, listening as deeply connected to the power of attention. This explicitly gives way to engaging a politics of recognition, and the dynamics of identity, social mobility, and institutional access, that includes contending with the affective economies at play within contemporary biocapitalism, which are always instrumentalizing the vitality of oneself as a situated body; not to mention the technological or medical approaches to "hearing ability," which draw out a bioacoustic politics. Through such an acoustic politics one may further address the issue of recognition by expanding understandings of language and voice to include the diversely abled, issues of translation and interpretation, and that attends to verbal and nonverbal, spoken and signed expression. Finally, by developing a framework of acoustic principles and practices one may better work on behalf of an acoustic commons, as the commoning that may articulate sound and listening as "social resources" in capturing a radical ethics of openness, a wild law.[39]

Acoustic justice, and the notion of acoustic principles, is understood to work at exercising one's freedom to listen,[40] as that which supports attending to the larger sphere of life, as well as the forces that act upon it, and that which grants an emotional and cognitive awareness as to those in need, those with or without voice, and those that also exceed the range of one's hearing: those I may never truly hear, but still, I know exist. The right to listen, as a counterpoint to the right to free speech, captures the necessity for turning hearing toward what must be heard: the expressions often occurring outside or beyond the acoustic norm of distributed sound. In this sense, acoustic justice lends to the forming of gestures and practices—listening practices, silent practices, careful practices—that attempt to reshape the arrangements enabling such rights and paths, and in doing so, modulate the norms by which we may encounter each other, and more. As a thick justice, acoustic justice is one that figures recognition equally as that which honors the unrecognizable as such.

Overview of Chapters

Acting as an expanded introduction to the question of acoustics, Chapter 1 sets out to map existing research and theories on the topic of aural architecture, as well as drawing key references to acoustic ecology and sonic effect theory. By bringing into play existing research and knowledges, the chapter moves toward deepening an understanding of acoustics as a political and performative framework. Through such a move, acoustics shifts from that of a property of architectural space to a critical means for elaborating questions of subjectivity, recognition, and social movement action and community struggle. Acoustics emerges as a question of the "distribution of the heard" in whose practice, from rhythmical organizing to vibrational worlding, communities may work at expressions of commoning, mutuality, and what I term "compositioning," as an improvisational, acoustic crafting in collaboration with the matters and energies of environments.

I elaborate this notion by reflecting upon a specific artistic project I developed in the context of the squat community, La Tabacalera, in Madrid (2017). Describing and narrating the project allows for grounding theories of acoustics within a particular context, and for moving forward by underscoring hearing and listening as not only questions of sound but also that of orientation, attention, and shared imagination. The chapter concludes with a reflection on the research collective Forensic Architecture, and in particular their project on the Saydnaya prison in Syria (in collaboration with Amnesty International). Their project captures the ways in which sound works as a form of violence, articulating acoustics as a technique of abuse. Through acoustic analysis and modeling, Forensic Architecture (with the artist Lawrence Abu Hamdan) provides an important view onto acoustic performativity, especially as contributing to justice work.

Extending from questions of acoustics, and acoustic performativity, Chapter 2 elaborates the concept of "acoustic relationality." This is based on reflecting upon hearing and listening as what enables encounters and collaborations with otherness: as a sensing of what is often unfamiliar and unrecognizable, yet no less proximate. Consequently, relationality shifts to forms of "extra-relationality," enabling strange forms of contact, a *being alongside* the nonhuman, the transmitted, or the unintelligible. I work at elaborating understandings of the extra-relational by addressing the field of sound art. Sound art is examined as a range of methods and practices that explicitly engage with the unfamiliar, the distant, and the foreign, in such a way as to position hearing at the limits of human sociality. Through a consideration of the artists and composers Alvin Lucier, Jana Winderen, John Grzinich, e-Xplo, Sarah Vanhee, Satch Hoyt, and Learning Site, the chapter leads to a speculative theory of resonance, or what I term "poetic ecologies of resonance." Such ecologies are underscored as manifesting acoustic relationality, figuring listening and the movements of resonant constructs as what may engage the unrecognizable. Within the dynamics of poetic ecologies, questions of recognition shift to that of resonance, which explicitly supports unstable and emergent forms of coexistence. The chapter moves further to question in what ways "listening consciousness" shifts under the intensities of resonance to that of "poetic dreaming": a "listening-dreaming" that enables one to navigate precisely between the familiar and the unfamiliar, the human and the nonhuman, the concrete and the spectral. Listening-dreaming is queried as a form of "dream-action" which, following Elena Loizidou, may contribute to negotiating the political complexity of globalization, enabling ways of trafficking between place and displacement.[41]

While Chapter 2 unpacks sound and acoustics as inciting an expanded relationality, an extra-relationality enabling modes of contact with the nonhuman, the foreign, and the strange, Chapter 3 brings focus back to the body and, in particular, to a question of skin. The skin is analyzed in order to capture sound as fundamentally a question of touch, of tactility and felt experience. Drawing upon the psychoanalytic theories of Didier Anzieu, and his concept of the skin-ego, I explore skin as an interpersonal material as well as a political question. From such a view, acoustics, and the ways in which it facilitates orientation, is drawn out by way of skin, accentuating the intimacies, the experiences of sensation, of inner touch, afforded by way of the vibratory, affective, and distributive qualities of sound and listening. Extending from Anzieu, the chapter draws into consideration the work of Monique Wittig, whose book *The Lesbian Body* is examined. In particular, I'm interested in how Wittig maps skin as what is stretched, strained, and shared between bodies, and by contending with the social (masculine) ordering of language, articulates a deeper, more erotic sense for the emancipatory potentiality of contact and consensuality. By way of skin, and the writings of Wittig, along with the work of Sara Ahmed, I pose the concept of "queer acoustics" in order to question in what way orientation is often a question of "straightening up" or "falling in line" with dominant norms.[42] Skin is subsequently positioned or unfolded as a "borderspace" following the theories of Bracha Ettinger, one that facilitates exchange and "shareability" rather than fortification.[43] In tension with a politics of borders, and the new regimes governing citizenship, borderspaces become thresholds by which an "art of negotiation" is performed, fueling other modes of touching and being touched. Finally, this is discussed by way of the Manus Recording project, an archive of audio recordings made by detainees held on Manus Island, Papua New Guinea.[44] Developed by curator Joel Stern from Liquid Architecture and scholar James Parker, in collaboration with André Dao, Michael Green, and Jon Tjhia, and participating detainees, the project draws out a compelling view onto questions of detention and bordering practices within today's geopolitical environment, and allows for questioning how skin and borders become central to new regimes of governing by way of citizenship (through elaborated modes of profiling and extra-territorial forms of detention). Listening into and along with the recordings, I argue for acoustic justice as a framework for contending with contemporary scenes of detention, figuring "skin-work" as a potential base for rethinking and reconfiguring understandings of community.

Throughout *Acoustic Justice*, I attempt to expand understandings of sound and listening by way of acoustics and the practices that work at reorientation, to shift from sound as strictly an object to states of attention and attunement, affording engagement with the social and political conditions shaping relationships. Chapter 4 draws this out further by specifically addressing cultures of the Deaf. D/deafness[45] is positioned not as a state of being without sound, but rather, as another mode of hearing, which may critically guide Sound studies toward a greater questioning as to what constitutes sound and listening. Such an approach is elaborated by addressing a number of key aspects of Deaf culture, including the visual-gestural language of Sign, articulations of Deaf space and poetics, histories of audism ("audio supremacy") and questions of diversity. Ideas of D/deaf hearing are thus mobilized in order to upset the normative view of sound as being "ear-based," leading to elaborated understandings of gesture and touch, seeing and speaking. D/deafness is further considered in order to bring into discussion the topic of voice: what constitutes "having a voice" within a framework of D/deafness? Reflections on voice by way of D/deafness are developed and put into dialogue with a range of theories so as to problematize dominant notions of verbal speech as the basis for social and political recognition. This allows for posing a notion of "gestural voice"—the understanding of voice as being shaped by signing, by way of embodied expression, and that supports articulations of cultural identity in diverse ways and directions. Finally, the chapter concludes by reflecting upon cochlear implants and the related discussions around bioethics and bioacoustic politics, which continue to affect Deaf communities. Considering the larger debates around neurodiversity and technologies of enhancement, I consider in what ways Deaf culture points out particular critical and creative routes that also productively impact onto Sound studies. If acoustics, as I suggest, underscores justice as a conflict over recognition, Deaf culture calls for a critical rethinking as to what it means to be heard.[46]

Freedom Is a Constant Articulation

In working at the concept of acoustic justice, I'm interested to attend to the ways in which social struggles often aim toward honoring a world of deeply rich entanglements, enacting "articulations" which, as Donna Haraway argues, counter representational politics.[47] Articulations, as Haraway suggests, allow us to stay close to the lives, the communities, the urgencies which politics often frames as representational, as objects of fact; articulations as fostering an immediacy of justice. This concern for articulation over representation, for myself, has found particular traction by engaging with the qualities and affordances of acoustic relations—a thinking relations acoustically. I have found it productive and helpful to approach the contemporary challenges around "life under Empire" by adhering to the experiences and potentialities I feel an acoustic model engenders.

Acoustic justice as a proposal extends its focus on articulations by elaborating sound and listening as dynamic pathways of sense, subjectivity, and collaborative living. By way of acoustic performativity, poetic ecologies of resonance, the complex permeability of bodies and borders alike, and finally, through the forms of listening and speaking articulated by way of Deaf culture, acoustic justice is mapped as what works at giving support to experiencing a diversity of relations—as a continual work of reorientation. Such a proposal, and body of work, I also intend as a contribution to Sound studies: to bring a more pronounced sense for how acoustic knowledge, and discourses on sound and listening, may engage questions of justice.

Bringing into focus ideas of acoustic justice, I'm also interested to further enrich the continual endeavors aimed at *relating otherwise*: otherwise to the systems of predatory capitalism that makes of one a subject shaped by the normativizing of a state of catastrophe—grounded in what Angela Davis emphasizes as toxic individualism.[48] The gestures and acts by which one may upset such interpellations, such systemic conditioning, are brought forward here as an acoustic question or challenge—an acoustic articulation; that the endeavors to decolonize the social and geopolitical orders, as well as the ways in which such orders effect the power of the body, are supported by following sound and listening as deeply transformative socio-material channels—poetic, slow channels that are enabling for community, especially versions that remain diverse, agile, strained by the reality of life itself, and that may stress and make responsive in return.

1

Acoustic Performativity

Practices of Composition

The work of architectural acoustics often sets out to refine or reform existing environments, focusing on noise, vibration, reverberation, and other auditory issues or disturbances. This is mostly developed through material interventions that set out to absorb or reflect particular frequencies or vibrations, thereby tuning an architecture or environment so as to better enhance the experience of particular sounds. Acoustics is therefore a work, a set of strategies, that contributes to defining the movements of sound through an environment, producing or supporting the emergence of particular sonic images. In this sense, acoustics may be thought through in order to unpack larger questions as to how hearing and listening, and the arena of sound, in general, are affected by various external forces and knowledges. To do so, I attempt to detail aspects of acoustics, moving from questions of architecture to aurality and the fostering of social life. This leads to an understanding of acoustics as a range of gestures and practices that support forms of physical and social orientation, and what Sara Ahmed terms "the work of reorientation."[1] I elaborate these perspectives by reflecting upon two different projects, one related to an artistic work I developed at La Tabacalera, Madrid, and the other produced by the research group Forensic Architecture, in collaboration with the artist Lawrence Abu Hamdan, on the Saydnaya prison in Syria. Following these examples and inquiries, acoustics is underscored as a performativity, a practice of composition embedded within struggles over recognition and justice.

Acoustics: The Distribution of the Heard

Along with defining or shaping the movements of sound, acoustics affects the passing of communicational and emotional feedback, and the capacity to feel oneself as part of an environment or social milieu—how feelings of belonging are influenced by acoustic densities and volumes, reverberations and noises. As such, acoustics can be understood to impact not only on the spaces we inhabit but, also on the affective contouring of private and public life, lending to experiences of place and belonging by conditioning who or what is heard, and whose voice may gain traction within particular environments. Belonging and recognition, attunement and well-being are thus deeply embedded within the question of acoustics, and the capturing of often "agitating" sounds acoustic design works to balance. In this sense, I'm considering acoustics as a question of "design"—a culture of sonic strategies, a science—as what contributes to the articulation of a sonic image and the figuring of sound within environments, as well as a broader more speculative framework for addressing issues of orientation and place-making. In this sense, I'm keen to underscore acoustics less as a property and more as a performativity; acoustics as a practice, as what we do to enable particular orientations, for listening to *take place*, to have a place. If listening can be understood more broadly as modes of attention, then acoustics is what we do to facilitate attention, and which also allows for thinking through what it means *to sound* and *to listen*. From sonic materiality to spatial and platial constructs to the affordances of hearing and being heard, acoustics is posed here in order to critically pursue the social and political matters that condition subjectivity along with the experiences of place, of *being placed*.

By considering these perspectives, acoustics emerges as a political question—a politics around which the urgencies of hearing and being heard are played out. In this sense, I'm aiming to position acoustics, following Jacques Rancière, as *the distribution of the heard* in whose practice or operation we may detect a range of contestations and imaginaries, each defining in various ways forms of orientation.[2] As the distribution of the heard, acoustics contributes to the ways in which we orient ourselves, or are oriented according to the sounds we are able to hear or sense, and how we may feel ourselves being heard—how one is situated within a greater acoustic ecology. While acoustics as a knowledge is often put to use by developers, engineers, technicians, and urban planners as a means to assist in ordering the built environment, regulating excessive reverberation and decibel levels, and organizing particular architectures, I understand acoustics additionally as an everyday practice—a range of gestures in which people actively modify or retune their environments to support the movement or flourishing of particular sounds.

Barry Blesser and Linda-Ruth Salter open up this perspective in their consideration of what they term "aural architecture." Aural architecture is theorized as being both a question of acoustics and the shaping of a given architecture, as well as sociability, where people also participate in defining an "acoustic arena." As they outline: "The acoustic arena is the experience of a social spatiality, where a listener is connected to the sound-producing activities of other individuals. By manipulating the spatial design, the aural architect influences the relationships among the occupants of a space in a multiplicity of acoustic arenas."[3] Importantly for Blesser and Salter, acoustic arenas are often manipulated through rather ordinary gestures, from opening and closing windows to adjusting home appliances; from listening to music to contending with the sounds of neighbors, building home cinemas or organizing parties, creating comfortable spots and crafting quiet areas often against the sounds one cannot control. Such overt gestures are also complemented by the more nuanced ways in which acoustic arenas are punctuated or attenuated by a general flow and rhythm of behaviors, from daily rituals to caring for the demands of others. In this regard, aural architecture integrates different acoustic registers, from the psychoacoustic to the electroacoustic, from the social acoustic to the bioacoustic; from the experiences of hearing to the diffusions of mediated sound, and the socialities and ecologies shaped and defining acoustic worlds. The ongoing movements of acoustic arenas thus come to impact onto the distribution of the heard, recomposing through a range of gestures the articulation of a sonic image—a "room tonality."

The issue of an acoustic politics, and the practices or acts by which such politics plays out, may be additionally captured by considering ideas of "fidelity" and "reflection," which are often at work in defining standards of acoustic design. From an acoustic perspective, fidelity can be understood as the resolution of a sonic image which brings focus or intensity onto particular sounds, giving clarity to a particular distribution or diffusion of sonic matter. Within the production of a sonic image, acoustic design often places emphasis on *staying true* to particular sounds, thereby tuning architectures, within concert hall design especially, so as to minimize disturbance or coloration of a sound. Fidelity is therefore often based upon normative agreements as to what constitutes "good sound" and by extension, distortion, or "dirty sound." Such agreements therefore contribute to placing value upon particular acoustic as well as aural behaviors on the part of a public.

Nina Sun Eidsheim provides insight into the establishing of "acoustic norms" by addressing the concert hall experience, and, in particular, how experimental forms of music and performance often draw out such normativity, leading us to recognize the often rigid parameters that define how and what we may hear. Importantly, Eidsheim highlights how acoustic norms greatly inform the establishment and maintenance of "acoustic communities" that form around particular musical cultures and concert halls along with their standards of acoustic mediation and reception.[4] As an example, she highlights how concert hall design within Western classical music traditions have tended toward a two-second reverberation time, which allows for a sense of "immersion" as well as "clarity."[5] Such standards wield great effect onto what an acoustic community may come to understand as "good sound" as well as what counts as noise, helping to define beforehand assumptions around the "meaning" of sound. "Here, being faithful entails such virtues as being in tune and conveying the a priori intent and meaning of a particular sound, composition, or musical-cultural tradition."[6]

Fidelity may be positioned in order to critically tussle with the ways in which acoustics contributes to defining "habits of listening" in which particular sounds appeal to normative codes and conduct. Within such scenes and constructs, noise is figured, following Tim Cresswell, as being "out of place"—a sound whose wayward itineraries and subsequent distortions interfere with music proper.[7] While such depictions are readily found by following concert hall design, and the cultures of classical music, I'm curious in what ways fidelity, and the question of acoustic norms, impress themselves onto a more general arena of everyday life and daily geographies. From scenes of domestic inhabitation to exchanges on the street, or even across the digital networks that define a great deal of social life, acoustic norms and habits of listening help shape discursive, social, and behavioral patterns. And importantly, what we come to hear, or more precisely, what sounds come to have a place and are therefore brought within reach. To stay true to certain sounds, as fidelity outlines, is to also demarcate particular limits onto what counts as listenable.

Alongside fidelity, as lending to the production of a sonic image, reflection and reverberation support the *propagation* of sound through a spatial environment. Within architectural design, the use of absorbent or reflective materials, for example, significantly plays a role in facilitating and modifying the ways in which sound moves within a given space. Reflection, in a deep sense, lends a certain life or vitality to particular sonic images—the reverberant responsiveness of architecture, in fact, is often what registers the *expressivity of life* within places. Reverberation (and echo) acts as a carrier of sound itself and the concomitant sense of vibrancy it imparts; within reverberant "dead" spaces, such as anechoic chambers, one immediately confronts reverberation as key to feelings of *being somewhere*, and how places come to life through sound: the "room tones" that register the general humming of place. As such, architectural acoustics often works at shaping, minimizing, or controlling reverberation, and, accordingly, significantly contributes to the sense of vitality itself—the tonalities that underpin the ongoing humdrum of life.

Yet, reflection, and the movements of echo, while giving way to the propagation of sounds, also often extend architecture beyond its physical borders (as is often the case, too much reverberation may overwhelm a particular sound source, hindering clarity and communication). As an example, Matteo Melioli highlights how the "reverberant geometries" of the Basilica di San Marco in Venice contribute to aural experiences by acoustically "dilating" the limits of the physical space, giving way to a "ghosted" perception. In this sense, reflection and reverberation may "distort" physical space by transgressing its limits, its fixed lines. In doing so, the life and vitality of these reverberant geometries makes one sense that sounds are coming from a range of directions, thereby dramatically "confusing" an understanding of the location of sound sources. Within the Basilica such reflective dynamics have been put to use within specific music compositions, which, as Melioli suggests, guide a listener into an echoic, reverberant musicality that seeks to extend the physical limits of architecture toward a sense of "infinity."[8]

The capacity for reflection and reverberation to extend or confuse architectural space, charging one's psychoacoustic imagination with a sense for the ghosted and the infinite, for an ever-expanding distortion, may also be considered within more social situations. The acoustic experiences of reverberation may lend to gestures aimed at enlarging a sense of place or belonging—a *sounding out*, a throwing of sound (of voice, of noise) by which to rework one's sense for being emplaced within given environments, thereby affording possibilities for destabilizing or enlivening the territorial contours of place, to work at reconfiguring the perceptual and spatial lines that fundamentally contribute to the distribution of the heard. Such reverberant enactments, while distorting or animating the architectures around us, may work at capturing a sense of participation through all that may come echoing back: to stir the rather fixed lines of space in order to gain a performative responsiveness from what it means to be somewhere.

Acoustic communities are therefore nurtured not only as a question of musical cultures and their halls of performance, but equally within the everyday environments in which fidelity and reflection are appropriated and reappropriated, captured as terms within struggles over belonging. Whether in the living rooms or the classrooms, the churches or the offices, the acoustic figuring of reverberant geometries are spatial and temporal channels by which to capture ways of inhabiting the everyday.

Reflection and the making of a sonic image are essential to the feedbacking of oneself and a given place, which affects the experience of collectivity and community so important to nurturing this identity one may become: to transgressing precisely the spatial terms that distinguish being in or out of place. Whether toward the infinite and the transcendent, or toward a horizon of togetherness and collective determination and pleasure, or even in the daily routines shaping or reshaping individual life, reflection and the casting of sound it supports contributes a profound sense for moving beyond the physical limitations of place as what is often setting limits or even "disciplining" one's potential flourishing. As such, reflection and fidelity may be captured as acoustic tools and knowledges, even weapons, by which communities work at definition, at extending precisely the horizon of possibility: to both define a particular sonic image, a configuration, a delineation, a musicality, while animating its given features, stirring geometries with acoustic potential: to make room for life itself.

Following notions of fidelity and reflection, and the issue of acoustic norms and communities, I may begin to unfold a range of questions: What acoustic forces and forms exist that enable one's voice to resound within certain rooms or environments, and to find place within a particular sonic image or arena? What acoustic decisions have been made to support one's sense for being able to stand up, a standing in front and a speaking forth, as well as for imagining possible futures? And further, what forms of acoustic economy are at play within a given architecture or environment, and that work at orienting us in certain ways and according to particular norms or values?

Acoustics, as I'm suggesting, is implicated in the shaping of sociability, which is greatly nurtured by how one may align with particular tonalities while distorting others, giving way to expressions of agreement and disagreement, harmony and discord. Here, we can appreciate the practices or gestures by which people rework the distribution of the heard, retuning environments and a dominant tonality, an acoustic norm, in order to hear or feel differently, and to gain traction for other voices: to demand a shift in the project of fidelity by dilating or ghosting the resolution of a sonic image, enabling the reflection and propagation of other concerns and desires.

Acoustic Practices: Modalities and Orientations

Acoustics, as I'm suggesting, allows for a consideration of the ways in which experiences of sound and listening contribute to how one negotiates social and cultural life and one's place within it. From such a view, acoustics deepens awareness of sound culture as an arena that often forces and facilitates encounters across dominant and marginalized communities: how aural architectures may also figure as political arenas, for not only giving volume to the particularities and positionalities of one's identity—to state oneself, or to be systematically instated by way of identity—but also estranging the normative arrangements that make identity count by *hearing differently*. What do I mean by "hearing differently"? To what degree can sound really matter when it comes to struggles over recognition and belonging? In what sense might acoustic justice contribute to notions of justice in general, and specifically within particular scenes of violence and inequality?

I'm moved to understand hearing, and the meaningfulness sound may carry, through a range of particular expressions, including scenes of communal gathering and social ritual, events of collective mourning as well as celebration, for instance, where the importance of hearing in particular ways is potently staged. I'm further moved to appreciate hearing, and the more pronounced forms of listening, by way of numerous pleasures and pains that press upon the body, from the joys and ecstasies of listening to the ways in which sounds may greatly annoy as well as brutally harm. Hearing and listening therefore contribute to physical and mental health, suggesting a greater auditory framework connected to well-being and care. From the material and relational dynamics of sound, which assist in giving shape to ways of living, to the support found through the vibrant touch of others and the shaping of intimacy—this listening that brings us closer—or the frustrations and rage arising from contending with the tonality of dominant political voices. Such experiences and auditory movements greatly influence the ways in which one may grasp potentialities for self-realization, leading to a critical sense for listening's relation to social and political institutions, whether by emphasizing the importance of hearing within judicial environments, where law proceeds by way of the verbalization and the hearing of argument, evidentiary truths, and the pronouncement of judgment, or in the struggles over recognition, and the articulation of rights, which find traction through the power of attention, of being heard, and the acknowledgments we give and are given. Or I may underscore those scenes in which care and repair are afforded by way of listening, not only to what is said, but equally to the withdrawn yet persistent presence of the understated, the repressed, the inarticulated: to bring a profound sensitivity to the unsaid or the unsayable, and to those scenes where it is impossible to speak. Accordingly, one may begin to outline a complex list pointing at the ways in which *hearing differently* seems to matter.

Acoustic justice is proposed as a critical framework for unpacking what is at stake in hearing differently, leading to an understanding of sound as having a great effect on scenes of social, ecological, collective, or personal struggle. Yet, I'm also concerned to use acoustics in order to extend understandings of hearing and listening as not always being oriented by sound itself; rather, might we pose hearing and listening as equally being about attention, affection, introspection, well-being, and recognition? Hearing and listening open a broader framework by which to attend to bodily and social life, our interdependencies, as well as how to critically and creatively navigate a social world and its limits. To say, "I hear you" is to confirm a great deal more than simply the reception of sound.

In my previous book, *Acoustic Territories*, I approach similar concerns by drawing out six specific auditory figures or practices, including echo, silence/noise, rhythm, vibration, feedback, and transmission. These allowed for a consideration of acoustics as a broader political question, whereby the modulations of particular volumes, or the intensification of certain reverberations or transmissions, for example, contribute to whom or what is heard and how, and how hearing differently becomes a site of cultural struggle and expressivity. Acoustic territoriality is cast as a tension, articulated in the ways in which people create acoustic arenas and horizons, and their own reverberant geometries, specifically negotiating the spatial borders, acoustic norms, and social orders that enframe subjectivity.

The conceptual rendering of these figures or practices may be further mobilized here, around the theme of acoustic justice, in order to position acoustics as central to engaging a politics of recognition and misrecognition, belonging and unbelonging. In this way, acoustics allows for processes of social engagement and imagination, where *sounding out* works to secure forms of attunement and communal support, acts of linking, binding, celebrating as well as breaking away—acoustics as a tensed, energetic volume around which sharing and participation, rapture and fragmentation are held. Through such a critical move, how we understand the physics of sound is found not so much on an expert knowledge of wave phenomenon, or through a detailed account of the nature of hearing, rather, acoustics may be figured as a field of everyday practices or gestures that contributes to the working through of relations: acoustic practices that negotiate the distribution of the heard and the establishment of auditory meanings. In this sense, I'm curious to consider in what ways acoustics may be drawn out in the nurturing of shared environments by supporting the work of care and compassion, as well as for mobilizing possibilities around social struggles that necessarily distort existing patterns of audibility or fidelity: to put into question particular notions of sonic value. And further, how acoustics may emphasize a broader experience, where listening extends beyond the hearing of sound and toward elaborated modes of attention and orientation, of being in place and with others.

Through vibration and rhythm, for example, we may deepen understanding for the ways in which people's lives are affected by the spatial contours and affordances of the built environment, leading to forms of attunement and synchronization, as well as disturbance and injury. Vibration and rhythm are spatial and temporal channels, affording the pacing of specific movements and routines, the interlinking with spatial structures or institutional systems, and the socialities shaped by the diffusion or contraction of affective forces—that which touches us—and that impact greatly upon one's sense for being emplaced or displaced, able or unable, recognized or misrecognized. Vibration and rhythm, therefore, figure as means for modulating social or institutional bonds, and how one may synchronize and feel connected. Additionally, the tensions between silence and noise, as what often force negotiations between neighbors and strangers, further reveal acoustics as a performativity by which individuals and communities absorb or deflect a given volume, shaping and reshaping, articulating and defending their own spaces and temporalities, as well as their own well-being. Consequently, the tonalities of places are amplified or distorted according to certain practices shaping the flows of affection and attachment, access and mobility.

Judith Butler's theories of performativity may be utilized here, to open a fuller view onto acoustics as an arena impacting the work of subjectivity. As Butler outlines, the performativity of the subject is predicated on a structure of power and discourse that precedes appearance, that gives one a name and through which one works at inhabiting the social arena of meaning.[9] Subsequently, subjectivity is highlighted as being always already determined by an exteriority, a social order that enables one to appear, that names oneself, while regulating one's entrance through mechanisms of power and meaning. As Butler argues, one is "called into being" through an incorporation into language, an interpellation that enables as well as delimits access to a greater arena of social or institutional life. Performativity thus names the ways in which the "I" expresses itself by way of a "citational chain," a lineage of repetitions or reiterations, highlighting an "original complicity" with power—to repeat the words properly, to take up residence.[10] I move my body as others do, along the particular lines presented and made available; I speak as language is put into the mouth to shape its structure, its voicing, and in so far as particular conditions enable it, those laid before me, by others, and by the discourses that ask of one to perform, that come to move one into speech and against which one moves.

From the call into being to the echo one then works to perform, to take up residence, subjectivity is fundamentally beholden to an exteriority—it is, in fact, constituted by a complex relationality. In this regard, I am never fully in possession of myself as a subject; rather, I answer back to this fundamental call, this interpellation, through an echo that beats back particular social and linguistic demands, or that searches for ways to rework the field of meaning—to continually inflect one's voice through particular cadences, vocalizations, and wording: a phraseology, a musicality even, in search of individuation or recognition.

Extending from Butler's thinking, I may return to the understanding of acoustics I'm pursuing here, to pose questions of performativity as key to acoustic orientation and the working through of social relations it comes to influence. Here, I understand acoustics as being central to a relational grammar—a deeply immersive and physical knowledge path as well as range of skills acquired through the ongoing experiences and expressions we have of sounding and listening, of being called into being and of answering back, of signaling and exchanging, navigating and negotiating the social processes and pressures that require of one a certain answerability, a figuring of one's emplacement, and the subsequent forms of legibility one must pose.

Following from this critical framework, acoustic performativity is fundamentally a question of orientation—how we *find our way* by drawing support from the aural and felt knowledges of acoustic experience, as well as through the communicational, organizational, and affective capacities of its performativity, from the silences and noises, rhythms and vibrations that continuously press our bodies and shape our environments. While this may seemingly give privilege to those who hear over those who cannot, I would emphasize that hearing be extended by highlighting acoustics as a deeply tactile, vibratory, and percussive arena, as well as an organizational process or capacity lending to the configuration of bodily and social orientations and posturing.

For example, rhythm is not only the beating of time within music, but equally a "compositional" question, articulating greater processes of synchronization, temporal spacing, a sense for alignments and misalignments within personal and public life: the rhythmic tracks by which moving through the world is articulated. In addition, vibration is an energetic force impacting onto hearing that is equally figured as experiences of "feeling," contributing dramatically to a sense of place and connection.

This expanded understanding of hearing I'm suggesting, from the deeply tactile to the temporal and motile, may be furthered by considering the experiences and conditions of vertigo. While vertigo may relate to a fear of heights, or anxiety over certain experiences, it is also often based on an inflammation of the inner ear and a disorder of the vestibular system through fluid build-up or the accumulation of calcium particles, for example.[11] Vertigo, and the vestibular system, which connects the inner ear and ocular perception and assists in bodily and spatial orientation, highlights the broader physiological and neurological operations of the ear and auditory nerve, revealing their connection to how we come to *balance* ourselves. Subsequently, it becomes important to consider how the ear does not always lead to experiences of hearing, but, rather, is situated within a larger spectrum of physiological and neurological, and by extension, psychological and social capacities and behaviors. In this regard, acoustics allows for a view onto questions of health and well-being, from the psychoacoustic to the bioacoustic. In drawing out this perspective, the ways in which listening is often called upon to describe ways of being attentive, as what extends hearing toward capacities of empathy and alertness, as an emotional and self-reflective channel, and as something that animates our social bonds, may be brought into fuller play. In doing so, listening emerges as the work we do in cultivating and nurturing bodily care and social or interpersonal balance. Listening and hearing are therefore greater activities or capacities that capture processes of attunement and affection, and that enable the work of disagreement and recognition, all of which begin to elaborate acoustics as a deeply relevant framework.[12]

Listening, in this sense, is not only to hear but also to attune and detune, balance and rebalance the forms and forces by which one is figured as well as participates in the figuring of others. Acoustics impacts not only ways of hearing and listening but, additionally, also how people may come to access their physical, social, or institutional environments, navigating the situations and systems that work to give order to who may appear and where. In this regard, acoustics defines a range of processes around which practices or gestures of bodily orientation and recuperation, cultural expressivity and negotiation, social navigation and construction may emerge. As such, acoustics is always already political and greatly contributes to the performativity Butler poses.

Although Butler outlines a process of subjectivation that circles around language and power, one that works through the project of appearance and the possibility of speech and identity, an acoustic framework draws out other views. Acoustics, rather, shapes the experience of recognition, as that moment of legibility or mutuality, of answerability and participation, through a range of embodied capacities and enactments that are not always bound to language but, rather, specifically work as means for upsetting or elaborating the orders of linguistic signification. From processes of attunement and transmission to acts of discord and polyphony, acoustic performativity is operative within the continual work of subjectivity by specifically enabling movements in and around language and legibility. Acoustics, rather, situates one within a process that reconditions representation and the shaping of identity by elaborating the frame of address and its (textual) figuring of the body through the more tactile resonances and ghosted reverberations that continuously affect one's balance and emplacement. Acoustics, in other words, explicitly captures an affectivity, a sensuality, a noise—a distortion, or dizziness even—that supplements the linguistic work of social ordering, driving forward any number of heated debates, feverish exchanges, agitated volumes, compassionate caresses, erotic passions, ghosted imaginings, and punctuated steps. The distribution of the heard is a politics forcefully bound to tactile and felt experience, the intensities of interpersonal contact and sharing, the transmission of memories and fantasies, and the ambient energies often affecting social and institutional environments.

The externality Butler highlights as shaping and impinging upon subjectivity is here experienced as an acoustic dynamic, an acoustic performativity, one that contributes to processes of orientation and recognition through reverberations and resonances, rhythms and transmissions—this *sounding out* of oneself and others, drawing into relation the near and the far, the familiar and the unfamiliar, and how the external social world absorbs or deflects one's own noise. Acoustics is fundamentally a relational materialization; it works explicitly between and across bodies and people, things and communities, facilitating a sensing of self and world that relates to the immediate and the actual, as well as the ghosted, the virtual, the vertiginous and the deeply temporal, and which contributes to the sense of place a vital dynamic of attunement and detunement. This echolocational tuning—this embodied back and forth, signaling and navigating, a resonant, socio-energetic figuring (*the vibrational matters and noisy situations through which my voice must pass*)—exposes an acoustic performativity that must be understood to critically affect subjectivity, processes of place-making, and community building, as well as imparting a range of felt knowledges and tactics by which recognition is articulated or negotiated.

In order to think through acoustics as an arena of everyday practices, as a process by which people work at recognition and orientation, through the spatial anxieties and social dramas, I want to return to the conceptual figures or practices of echo, rhythm, vibration, feedback, and transmission that I considered in my previous work, in order to elaborate questions of acoustic performativity. By drawing out their particular qualities and dynamics, I suggest that such figures—material figures, poetic figures—may provide important channels by which one may negotiate the order of hearing, and the related acoustic norms that contribute to shaping oneself as a social subject. (I map these here as a lexicon that I will then elaborate through reflecting upon two projects in the final sections of the chapter.)

Vibration: As an energy, vibration passes through bodies and things, subjects and objects; it is penetrative and deeply connective. Subsequently, vibration collapses distance in favor of tactile contact, bringing things together within a material, resonant construct or scene of touch. Hence, vibration may be appreciated not only as a particular form of energy passing through matter but also one that may be deployed to generate forms of consensuality and commonality: an affective vibrancy and connectivity underpinning being together. In this sense, vibration provides an energetic and figurative means for conducting forms of sharing, even delirium, granting a tangibility to what we may hold or produce together. Vibration may be understood as an "ecology of feeling," an undulation of sense and matter that all places come to contain and that impacts one's capacity to belong or to contribute to a given environment: one enters, one senses the environment, this scene whose vibrancy grants support to feeling connected or cuts into one's ability to find comfort or balance. As an ecology of feeling, vibration may act as a base from which the "graspability" of things is articulated: vibration equally figures a politics of touch, capturing the ways in which things become available to us, or how one's own reach is delimited within a larger configuration. Vibration thus enables an extension of oneself, defining the limits or dynamics of being-in-touch. Whereas vibration is often disturbing to acoustic design, and the construction of acoustic environments, it equally imparts a potentiality for elaborating social contact beyond the lens of a certain legibility, occupying a territorial strata often situated below the line of appearance: an atmosphere, an ambience, a sensuality. In passing over and across bodies and things, vibration engenders a state of intense relationality, a being-in-touch that buttresses and exceeds identity—how often we are driven beyond ourselves under the joy of a certain vibratory pleasure! A scene of enraptured dancing shaped by the vibrational intensities of an electroacoustic amplification for example—that is, an arena of commingling often driven by what we term "a good vibe." Vibration, in this regard, is fundamentally connected to hospitality: the making of an ecology of feeling, the making possible a certain graspability, by which to welcome others: to incite and enrich "intermaterial" forms of common good.[13] This vibrational dynamic though is also captured and instrumentalized by a range of systems and initiatives in order to steer us in particular ways: to direct behavior, and the sensual and libidinal economies of desire, by way of vibrational, atmospheric constructs or mechanisms: the micropolitical capturing and instrumentalizing of affect. Accordingly, vibration may lead to a deeply erotic force of togetherness, this sharing into which I lose or regain myself, while also enabling forms of regulation.[14]

Echo: In contrast to the vibrational "commoning" of felt togetherness, echo may be considered as the basis for the crafting of difference, as a process of differentiation. Echo finds particular dynamic in relation to language performativity and the work of the voice. As the acquisition of language is essentially achieved through repetition, where one is called into a scene of repeating words as specific sound patterns, echo is potently figured—is outright mobilized—within the act of speech: to echo the particular vocal shapes of a certain language or linguistic home. The phonetic sounding of speech is therefore a framework in which we may learn of the powerful and complex performativity of echoing, where meaning and subjectivity are drawn out through the modulation of sounds and their repeated articulation. This includes a recognition of oneself, as a particular type of voice, a particular accent or ability that is always contributing to how one is heard. As such, echoing works to craft spaces or bodies of difference: we may echo each other in moments of social encounter, picking up words and returning them, reinforcing existing views, yet such exchange is central to outlining more clearly one from the other, to knowing of my body as different from another. The echo might be heard to draw out a tension around the notion of origin, enabling one to appear, to take up residence, while at the same time, granting a form of migration, a passage across territories of the familiar (family, nation)—or by also forcing one out according to the normative lines of "proper speech": this speech that can't quite fit in. This can also be applied to more spatial and architectural scenes, in which echo functions to find place, to echo-locate; throwing sound into the environment, through the social arena, one may be guided by all that returns: the reverberant tonalities and geometries that emerge and grant a sense of place-making. By extending a sound, a gesture, a voice—this echo that carries one through the world—echoing may generate a critical and creative framework for tensing origin in support of differentiation and narratives of passage, transition, exit: to capture origin as what we make of it. In this way, echo works in and around fidelity, a being *(un)faithful* to the original signifier. As such, echoing may reinforce as well as disturb, acting as a repetition that crafts possible trajectories and forms of orientation.

Vibration and echo begin to open the topic of acoustic performativity I'm posing here; as processes and practices, they suggest in what ways acoustics is active in facilitating the material shaping of places as well as that of subjectivity and the relationships that lend it identity. As tactile, relational capacities, the vibratory and echoic matters and movements that radically inflect and fill spaces between people, and that reflect across and around things, are suggestive for capturing acoustic justice, from the commoning of particular situations and scenes to the differentiating articulations and amplifications posed against or for others.

Feedback: I may further explore the possibilities of acoustic practice and performativity by considering feedback, which is positioned, following Barry Truax, as a communicational linking between people and environments.[15] This extends from expressions of inner voice, a speaking to oneself, a feedbacking by way of the acoustics of the inner self, to the sharing of speech and other forms of verbal and nonverbal communication central to fostering relationships. Communication is thus held delicately and dynamically within a framework of feedback, supporting a sense for being heard and for being in the world, and the need to sense oneself being sensed. I'm interested to emphasize feedback then as an unsteady, oscillating link—a performative, trembling thread of exchange continually tensed and modulated by input and output, by processes and experiences of interaction. There is a conflicted and highly sensitive quality to feedback, whereby a diverse range of voices vie for articulated presence—a *being heard* or acknowledged that must be understood to include both human and nonhuman beings, technologies and networks, material and institutional forces and systems. Feedback, and the communicational linkages it manifests, captures echoes and vibrations within a mesh of exchanges, alliances, and media, figuring an affective and informational movement and modulation of the sensible: of what can be perceived and folded into the sphere of shared meaning. As Truax suggests, "sound creates and modifies relationships, and therefore, when one relationship changes, so do many others, directly or indirectly."[16] In this way, questions of feedback can be extended beyond the open communications between people and their environments; rather, the oscillations of feedback radically index the availability or not of particular matters and spaces, and are captured so as to allow or to thwart access to institutional structures and social mobilities. Feedback may be a type of acoustic hinge, a flexed "informational" performative tensed by the pressures and possibilities of access, recognition, exchange, and the flows of communication defining private and public spheres.

Rhythm: Along with the acoustic dynamics of vibration, echo, and feedback, rhythm leads to questions of spatial and social relations, where personal flows and patterns work at experiences of belonging and assurance, rest and recuperation. The delineations of time, and the capacity to enter and exit spaces, to reshape one's immediate surroundings, can be appreciated as rhythmical formations and modulations of patterns. Rhythm, in other words, leads to issues of organization and power, how individuals and systems (social, institutional) align or misalign, or grate against each other, producing in their frictional meeting a pattern—a beating, a tempo by which we gain or challenge synchronization, a falling in line or a breaking away. Within the meeting of different forms or figures, different bodies and forces, rhythm emerges as a performative channel by which to literally *move* across or against time and space; a movement that figures into a construct of organization, an order—a repetition, a ritual even, a coming and going, which may work at the securing of trust: that sense for what one may rely upon, what, in fact, may return or is steadily there, and that I may synchronize with. In this regard, rhythm is always a question of the flows of power. As Roland Barthes suggests, the first thing power does is establish a rhythm.[17] Rhythm may force us into particular routines and orders; it may also be tuned in building relationships through shared movements, through a reliance upon what may return, or what one may call upon in moments of distress. As such, rhythm is the means by which to establish a certain time signature, a temporal frame for capturing a sense of purposefulness and direction. The urban theorist

Stavros Stavrides highlights rhythm as "creative responses to violent ruptures of time," which enables "a constant negotiation with experiences of temporal discontinuity" and processes of exchange.[18] Rhythm thus works at a structuring of time and place, a channel of negotiation, for setting pace to one's own needs or hopes; it articulates a scaffolding across which bodily, social, and political life work as organized forms, lending to configuring structures of support and freedom.

Transmission: As a final form of acoustic practice, transmission may be posed as a means by which one may *project* onto the world another set of meanings and messages, figuring within a given environment a range of possibilities and interferences. Transmission, in short, is a deeply complex process of *imagining otherwise*. An emanation, an otherworldliness, a projection, a passing down, transmission introduces a radical articulation of strangeness into our relational imagination and experiences: in passing down, whether in the dissemination of knowledge, or in the broadcast of information, transmission is always coming from somewhere else; it is a carrier, an intruder. Transmission is also reliant upon a medium of use, a technology, a technique, and as such, it is prone to breakdown, noise, mishearing. Along the way, transmission may err; it may overstep. In this sense, I'm interested in transmission as that which affords a deep sense for interruption: by this signal that invades the scene, supporting or instigating relations with the strange, with otherness, with voices or matter always out of place, whether from a distant location or in the passing down of family memories that are not directly one's own. Transmission often aligns with the unconscious, with the depths of knowability, whether in the transport of the radically formless and infective, or in the sharing of narratives of the past. Transmission becomes an important carrier, and is therefore open to forms of contamination, interference, static, and the figuring of a mutational future of recovery, connection, and fragmentation: the making of a rather unstable community. In this way, I understand transmission as a framework in support of all types of monstering, whether in the projection of fear of others or in the hopes for contact with what lies beyond human intelligibility. It thus provides, on a more banal level, an extremely operative means for figuring escape routes, and for evoking another set of social principles and possibilities: to capture the signal in favor of other messages. Here, transmission is also instrumentalized, again, by opposing forces and desires: by the logic of "carrying messages" it is prone to both dominant and minor expressions, from state administration to pirate acts, from national identities to counterpublic expressions. Transmission is always poised around different worldviews and agendas.

* * *

In detailing the acoustic figures or practices of vibration, echo, rhythm, feedback, and transmission, I intend at opening understandings of acoustics as a deeply effective base for critically and creatively approaching what's at stake in listening and hearing. The material and social conditions of places, and the articulations and expressions they may come to condone or corral, prompt or require, are brought into relief through acoustic interventions and arrangements, from the electroacoustic to the social acoustic. Acoustic practices or gestures may be captured as socio-material frameworks through which agreements and disagreements are made, communities assemble and disassemble, and the general flows of sonic warfare or acoustic welfare are channeled.[19] What they start to map out, as a type of topography of the sensible, are the ways in which performativity is routed through the felt knowledges and sonic experiences found in being a subject in specific places—a body whose capacities for engagement are greatly shaped, delimited, or supported according to the material and affective economies always at play: economies of power and discourse, as well as those of attention and recognition, sense and nonsense. In this regard, acoustic practices or gestures open up vital routes—a vibrancy, a sensitivity, an intelligence—for modulating and accentuating relational dynamics, and that may be understood to underpin struggles over recognition.

Ghosts, Distortions, Frequencies

The acoustic figures and practices I'm detailing here capture acoustics as a process by which to shape or interrupt a given order of hearing, thereby elaborating a framework of recognition and legibility. Acoustic performativity is a thick mesh of oscillations and frequencies, pressures and agitations that allow for flexing, detouring, fortifying, or relaxing one's sense of place. This extends into the ways in which we may refigure ourselves, reworking the synchronizations and alignments our bodies are often fitted to through more energetic, animate, affective, and percussive forces and actions: *refusals that cut into the patterning of the world.*

I highlight acoustic practices as wielding forms of knowledge as well as weaponry, material, technical, sensual, poetic weaponry that may afford passage across a diversity of situations and settings, where silence, for example, may support the transport of a certain message, or echoes work to enable the crafting of a differentiated space for marginalized or disobedient bodies. The spatial and social contours of the built environment are detuned and retuned, shaken and rattled, beaten in and out of shape, through a range of expressions and interferences that lead to a reworking of the heard and the felt as a project of acoustic justice—the gestures enacted in order to support the movement of other voices and other wishes.

Acoustics contributes to defining ways of hearing and being heard, as well as modes of attention and orientation, from the whispered word to the shouts that interrupt, movements of sound, of voicing, are always working through a given environment and its tonalities. As a fundamentally spatial act, acoustics dramatically influences the life of aural architectures, and yet acoustics is never so stable or clearly marked. The movements of sound are by nature hard to control, and sound mostly challenges compartmentalization and categorization—it is fundamentally a mix of frequencies and sources, projections and propagations. As such, sound is deeply impactful onto space, and as Matteo Melioli poses, contributes to forms of spatial geometry a range of animate elements that often "distort" and "extend" the physical delimitations of architectural space—leading to what Melioli terms "the ghosted and the virtual."[20] Through an array of movements, reflections, and volumes, sound fills an architecture, flowing across thresholds or over partitions; it may make the floor or walls tremble with a vibrational force; it may reflect across surfaces, wrapping itself around columns or ducking into corners, to soften the hard edges of the built, blurring distinctiveness or immersing one within sudden clarity—the articulation of a presence, a need, a wish. Such acoustic movements and projections are dynamically shaped by architecture, and yet, at the same time, emerge as an architecture unto itself—an *aural architecture*, a virtual space, a ghosted arena, a material ambience.

By following the relations between space and sound, we can appreciate how architecture comes to produce specific acoustic or reverberant geometries—propagations and projections, echoes and vibrations—that both fulfill and extend, upset or distort their architectural counterparts. Acoustic geometries are thus particular and can be appreciated in how they situate listeners within aural architectures which, as Blesser and Salter outline, result from the intersection of architectural space and the movements of sonic events. "Although we usually think of a soundscape as a collection of sonic events, it also includes the aural architecture of the environment. The experience of listening to a sermon in a cathedral is a combination of the minister's passionate articulation and spatial reverberation."[21] Aural architectures are thus dynamic formations constituted by the materiality of a built space and the subsequent reverberant energies of acoustic propagations. Yet, aural architectures, as Blesser and Salter pose, are additionally shaped by the interpretations people bring to what they hear, emphasizing the architectural experience over the physical geometries of the built. Here, architectural experience includes not only the articulation of a certain audible event but also the reverberations and vibrations that touch the body, that divert one's gaze, that haunt our attention—causing joy as well as anxiety. As such, aural architecture is not strictly audible; rather, it integrates visual and tactile qualities, as well as their rather virtual or emergent undercurrents, which may underscore acoustics not only as a physical architectural property, or a set of material or technical decisions, but also more as figuring a complex ecology of affect, imagination, corporeality, experience, and understanding: a multimodal and multifaceted "climate."[22]

As previously discussed, the making of a particular sonic image entails a working through of the physical architectures that condition the movement of sound—the interweave of the audible event and its reverberant spatial diffusion. Yet, I want to emphasize an additional aspect when considering acoustics: the phenomenological dialogue taking place between sound and architecture and, further, how the sounds one hears are also constituted by an array of frequencies, each of which contributes to the movement and comprehensibility of sound within an environment. This perspective may help in exploring the questions of sociality and recognition I'm pursuing, and in what ways acoustics, and an acoustic performativity, contributes particular input to defining place.

To enter into acoustics as a general framework is to also confront the specific "voices" that reside within any given sound. In this sense, we may speak of sound as both an extrinsic and intrinsic event, a singular object (i.e., barking dog) constituted by an array, a multiplicity (as a composite of frequencies). One's sensory perception of sound is a skillful balancing of these aspects, allowing one to register the extrinsic event (barking dog) within a given environment (or aural architecture), while understanding the qualities of that event, what particular features (timbre, dynamics, emotionality) that singular object contains and expresses.

Trevor Wishart elaborates this view by emphasizing how sounds are often working against each other, leading to forms of "masking" that makes it difficult to recognize precisely the source or nature of a sound. For instance, "loud sounds in particular frequency bands may prevent us from hearing those frequencies as constituents of another sound."[23] This concern has also been taken up within the field of acoustic ecology, and the study of environmental sound, to indicate the ways in which animal life, for example, may be threatened by introducing particular man-made sounds. For example, the sounds of automobile traffic produced along a stretch of highway may impact onto species of birds by specifically occupying the range of frequencies used in their communication. Interferences within the distribution of sound may therefore endanger the balance of ecological life by overpowering the movement and reach of a particular bird song, for example. R. Murray Schafer captures this point through his notion of "hi-fi" and "lo-fi" environments. As Björn Hellström summarizes: "Hi-fi designates an environment in which discreet sounds may be heard clearly because of the low ambient noise level. Lo-fi, on the other hand, designates an environment in which the silent sounds cannot be heard; the discreet sounds are masked because of overcrowding and/or high ambient noise levels."[24]

The distribution of the heard is a distribution of "voices" that contribute to shaping sociality, and the communicational feedbacking active within finding place. Yet, such distribution is also a distribution of frequencies constituting those voices, in which one sound may overpower another, shifting the qualities of a single voice by disturbing or encroaching upon the specific territorial line or bandwidth within a given frequency range.

I highlight this aspect of acoustics in order to deepen the understanding not only of the architectural experience but also the particular dynamics active in the movement of sound. Hearing and being heard take place as important events vital to being with others, yet these events are always conditioned by the particularities of place itself. Acoustics, in this regard, requires an additional understanding as to what constitutes place, for place is not only the room in which we meet or gather, or the voices heard therein, but equally the territory of the frequency spectrum itself and its distribution, its intrinsic ecology. Acoustically we live within an external, spatial arena alive with sound, and yet that life is taking place through a movement of frequencies, tonalities, micropatterns, each of which impacts onto that spatial, social arena. One's voice comes to occupy the room as well as a frequency range within the distribution of the heard. This can be appreciated by recognizing the ways in which masculine voices have come to wield a certain "authority" by sounding deeper, stronger, which necessarily overpowers or masks other, more high-pitched voices. Its place within a band of lower frequencies carries a profound impact, accentuating or reinforcing an acoustic norm that registers deep sound with power. To challenge authority, for instance, then may require accessing or distorting, on an affective and acoustic level, the frequencies by which certain voices often take charge, to demand a shift in the acoustic norm by which power resonates.

Acoustic practices therefore enact forms of retuning and modulation, interference and interruption precisely across these territories, this complex ecology; they mobilize a range of sonic, tactile, and reverberant forces, often drawing from the geometries figured by way of acoustic propagations and experiences, to support other configurations of power and discourse. As such, questions of recognition and legibility must take on, from an acoustic perspective, not only the enactments of affection and attunement afforded by way of listening but also the rather invisible conflicts waged by way of frequencies.

Acousticking and Compositioning: A Sticky Theory

I've been working through a set of perspectives onto acoustics in order to articulate a politics of sound and listening. Questions of performativity, and the gestures or practices that capture or negotiate acoustic norms, underscore how the distribution of the heard impacts onto individual and collective orientations. What might such practices, and the general work of reorientation, suggest in terms of acoustic knowledge, as a particular way of knowing? And what might acoustic knowledge bring to materializing forms of intervention onto situations of social struggle?

In his seminal book, *The Soundscape*, R. Murray Schafer underscores the question of lo-fi and hi-fi environments as central to what he terms "acoustic design." Here, Schafer highlights acoustic design as a profession that may work to rebalance existing "lo-fi" environments, seeking to tune the distribution of frequencies so as to enable one to "learn how sounds may be rearranged so that all possible types may be heard to advantage."[25] In mapping acoustic design, Schafer comes to describe "the soundscape of the world as a huge musical composition," locating us as "simultaneously its audience, its performers and its composers."[26] Consequently, acoustic design is posited as a work of "orchestration" by which an extremely active tuning of the world may intervene to give room to a plurality of sonorous life. This would include conserving important "soundmarks" while enabling a deeper appreciation for "positive silences"—silences that may work to recalibrate the soundscape and our place within it.[27] Acoustic design is therefore not only posed as a practical profession focused on sonic materiality, on defining and preserving sonic images, or even social questions, but additionally one involved in a concern for physical and ecological well-being. This leads Schafer to ultimately question the way acoustic design may orchestrate this musical composition of the world: "Which sounds do we want to preserve, encourage, multiply? When we know this, the boring or destructive sounds will become conspicuous enough and we will know why we must eliminate them."[28]

Although Schafer's proposal of acoustic design opens up the question of listening, and the importance of the soundscape in general, I'm concerned to upset the general assumption that "boring or destructive sounds will become conspicuous enough." In contrast, I'm focused on raising questions as to how we understand well-being, acoustic betterment, and the orchestrational project an acoustic designer may perform. In short, to shift from notions of acoustic design to acoustic practices of the everyday, which may help to remind how soundscapes are deeply territorialized and contested, claimed and reclaimed by a range of communities that may upset notions of "improvement" with an insistence on particular noises and non-normative listening habits. The ways in which one seeks or fights for orientation through a range of acoustic gestures and constructs, attunements and interruptions, and through a diversity of situational and ethical volumes or transmissions, highlight acoustics as a "sticky" matter.[29] Here, the movements of sound leave their traces; they impact onto architectures, leaving a residue as they go, in echoes that trail behind and that spread their ghosted repetitions, or through rhythms that punctuate certain escape routes. It may be better then to speak of acoustics, as I'm posing, through the concept of *acousticking*, a sticky performativity that works at time and space, the here and now, energizing materiality by vibrating molecules, by heating up any static notion of identity.[30] Sounds stick to us even as they pass, effecting one's sense of place and belonging, throwing us off balance; within the flows and force of their movement, sounds unglue us, trembling and exciting the matters of the body and its participation within any number of social configurations, time signatures, or spatial parameters. As such, we may also glue things back together by way of sound, reconfiguring their arrangements through a range of acoustic acts.

Moving from notions of acoustic design and orchestration, I'm interested in how acoustic practices become performative through such sticky processes and pressures, enlivening or distorting the articulations of space with emergent geometries and reverberations, which provide a sense for the sticky communities that also emerge from out of such movements of sound, that stick together precisely by way of shared orientations: communities that *acoustick* a set of meanings for themselves.

Acousticking works at forms of social navigation, bodily orientation, and cultural expressivity by way of acoustics and its performative potentiality. In doing so, I'm concerned to think through acousticking as something that may shift from Schafer's notion of "orchestration" toward what I may call *compositioning*. Compositioning is drawn out here to consider how acoustic practices, and the acousticking enacted, works at orientation—how one navigates a larger spatial and social order in search of a certain figuring, an amplification, a manner of living, and how orientation is constructed or fought for through an acoustic performativity: a vibrational commoning, an echolocational differentiation, a rhythmical reorganization, a transmitted contagion, a noisy displacement, all of which do not so much orchestrate the field of audibility, or compose from a distance, but rather, contend with the overlapping temporalities and spatialities, as well as frequencies and oscillations that emplace us, that stick to us, or become sticky and therefore malleable, and that we seek to make meaningful. Compositioning is an embedded process that grabs what it can, and that arranges along the way.

The work taking place at Cresson, a research center on sound and urbanism at the National School of Architecture in Grenoble, additionally considers the complexity at play within acoustics. Initiated by the social philosopher Jean-François Augoyard, research at Cresson has concentrated on "sonic experience" and, like Blesser and Salter, argues for understanding sound as always already contextual and highly situated. This leads to developing complex understandings of sonic experiences by integrating analysis of particular sound events along with musical, architectural, psychological, and sociological speculation and theorization. Sonic experience is thus never only a question of sound, nor the particular physics at play in acoustics, but rather, the emergence of a set of "sonic effects." Sonic effects, according to the researchers at Cresson, are the result of the interactions between a sound event, its context, and the experiences of listening, and which allow for mobilizing a larger set of theoretical tools. As such, sonic effects are captured by way of objective fact, discursive reflection, and subjective experience and interpretation. This is to be taken as a highly useful analytical platform, whereby the sonic effect "gathers together into a unified and harmonious listening what other disciplinary knowledge divides."[31]

In following notions of sonic effect, which is given in-depth consideration in the related book *Sonic Experience: A Guide to Everyday Sounds*, listening acts as a central thread, yet, at the same time, does not necessarily emerge in a unified and harmonious manner; rather, through a persistent categorization, sonic effects are analyzed by way of the separate fields of psychology, applied acoustics, sociology, musical aesthetics, architecture, and media expressions. Such interdisciplinary perspectives open toward understanding sonic experience as extremely dynamic and are suggestive for capturing sonic knowledge as something that applies to a range of disciplines. At the same time, I'm concerned that such an approach falls short in following through with what sonic experience seems to pose, as a deeply antagonistic upset of categorization. In this regard, sonic experience may also demand a more robust or speculative approach in terms of understanding how, in moving across such a range of disciplines, it also pulls them together, interferes with their categorizations, or proliferates their meanings. If sound and sonic experience can contribute to a broad field of knowledges, it may require a particular discursive approach, one that is more *transversal*, and that disregards the differences one might make between psychology and architecture, sociology and musical aesthetics, or underscores such differences as the basis for a more disjunctive science. Sound studies and its discourses, from my perspective, are an opportunity for staying close to sonic experience and its effects as a *transversality*: whose dynamic force, in making links between knowledges, also cuts across their demarcations. In this sense, notions of acousticking, and the work of compositioning, are posed in order to capture this transversality, the tensions and viscosity at play in sonic experience, and that one may draw upon to rework the categories that often emplace and orient in specific ways, whether on the street or in the university.

Consequently, listening may unify, but it does so by explicitly guiding toward a particular paradigm in which knowledge is embedded within the tactile, figurative, and equally disorienting flux of acoustic becoming. While we may work at forms of orientation within the reverberant geometries propagated by sound and sonic experience, such work is always already in dialogue with the ghosted and the virtual, the transmitted—that which is there and not there at the same time—as Melioli reminds, and which the urban sound designer Caroline Claus further underscores as mobilizing a "multiplicity of trajectories."[32] The particularities of acoustic climates, of aural architectures, must be appreciated as echoic, vibrational, and noisy experiences that coalesce and conflict over time, in the percussive, reverberant, and at times, quiet rush of life. As such, one must tend toward a fuller *poetic* approach in order to follow the rather creative science sonic experience seems to necessitate. I understand poetics, in this sense, following Donna Haraway's notion of "diffraction"—as a mapping of interference: an approach not of "replication, reflection or reproduction," but rather of generation—a generative tendency that necessarily exceeds itself, that proliferates.[33] A poetic approach to acoustics (and its analysis) might capture the generative complexity of sonic experience as an extremely diffracted pathway of feeling and knowing, sensing and doing that is also not without its anxieties or traumas.

By adopting a more diffracted and poetic perspective, I'm looking to emphasize acoustic or sonic knowledge as a map of interference, a mapping of the interruptions and transversality sound and listening seem to deliver, and to make possible. A poetic approach—a poetic science—might contribute to understanding the "multiplicity of trajectories" engendered by way of sonic experience, and an acoustic model, affording an investigation that stays close to the rather sticky movements such trajectories enable or impart. From the deeply felt passions of vibrational pleasures to the reverberant geometries that remain rather unstable, sonic experiences, and the aural architectures they produce, are highly energetic and, accordingly, are suggestive for a generative, diffracted discourse, a sound study (and even, an acoustic design) that would remain rather undisciplined and vertiginous.

Vertigo, in this sense, may not only be an anomaly within experiences of orientation, a disorder of the inner ear and its connection to the brain, but also, an aspect to hearing itself: a symptom of hearing differently, or that emerges in response to the orientations that would have us straighten up. Vertigo, as I'm suggesting, may educate on processes of orientation that are equally informed by disorientation, interference, supplying acoustic or sonic knowledge with a sense of disequilibrium or dizziness and its productive power. Acoustic orientation, following the generative and diffracted tendencies of sonic experience, is thus equally touched by disorientation, and in this regard, may assist in capturing ways of knowing as well as ways of navigating through the orientations and categories that often seek to territorialize: that reinforce disciplinary lines. The reverberant geometries that figure within our environments, ghosted by an emergent sense for a multiplicity of trajectories, from cathedrals to dance clubs to classrooms, put us in touch with a general auditory potentiality, one that may articulate epistemologies radically traversed by sensuality and imagination, by diffraction and proliferation: by what we hear and where that hearing takes us, thereby helping us to map the interferences that constitute life in the making.

Acoustic Commons: Sound as a Social Resource

Acoustics is posed as a critical and creative framework, one that can enable research onto sound and listening, architecture and embodiment, as well as approaches to ecological entanglements and challenges around social recognition. Such an arena of study, as I'm working at here, may enrich understandings of sound as being an acoustic object, an acoustic figuring of spatial and temporal distributions, a phenomenology, as well as allowing for engaging social and cultural life—a social acoustics that lends to a politics of recognition or belonging. As I've argued, this includes a consideration of acoustic practices that work at forms of orientation or balance, and that move us across the actual and the virtual, from acoustic norms and the power of frequencies to the acousticking of certain interruptions. In articulating such an approach, I want to conclude by reflecting on two projects. While taking place in very different settings, and with different purposes, they allow for a greater probing of acoustics as well as understandings of acoustic justice. This will further lead to a notion of acoustic commons, exploring acoustics as participating in movements of free culture, cooperative spaces, and participatory community models. Such expressions of "commoning" are reflected upon by considering how acoustics, and the related practices and knowledges it supports, figures within the making of common life: this being-in-common *compositioned* by way of acoustics and its affordances. In this regard, sound and listening, as capacities for attunement and interruption, may be captured as social resources.

In his book, *Experimental Practice*, Dimitris Papadopoulos reflects upon the ways in which individuals and communities of struggle work to transform the conditions of their own sociopolitical worlds, or what he terms a "region of objectivity."[34] For Papadopoulos, such work often exists prior to any self-identified articulation of social movement or oppositional politics; rather, the emergent gestures and actions that articulate positions or forces of challenge perform across and within everyday life to give way to the mobilization of shared resources, the crafting of informal networks of support, and the making of self-organized tools and resistances—gestures and actions that intervene within a region of objectivity to resituate the institutional structures or environments that limit possibilities for change. Through his argumentation, Papadopoulos opens up an insightful view onto social movement action, encouraging a broader engagement with the often less visible or outspoken demonstrations—the "imperceptible politics"[35] that come to radically influence struggles over social transformation.

In questioning social movement action, and what constitutes its productive force, imperceptible politics evoke another form of social movement, and the ways in which struggles over direct participation and emancipatory agency gain traction by often carving out daily formations of support and action—bringing attention to social care and material crafting. Through the notion of "practices of composition," Papadopoulos ultimately underscores how such modulations and intrusions onto existing sociopolitical conditions work as emergent constructs, which stake out a claim onto legal, social, and political ordering. Practices of composition are thus not just socially constitutive, but rather work to "ontologically" organize, building from within a given context or environment a possible world. Practices of composition attempt to shift the very terms and conditions by which particular struggles or conflicts are inscribed and shaped. As a result, individuals and communities of struggle forge possibilities for change by recomposing what passes as rightful, legal, or meaningful, giving way to what Papadopoulos terms "more-than-social movements": "More-than-social movements do not attempt to contest power by organizing protest; rather, they attempt to create the conditions for the articulation of alternative imaginaries and alternative practices that bypass instituted power and generate alternative modes of existence."[36]

I'm interested in Papadopoulos' argument, and would like to situate his ideas here in order to further articulate the ways in which acoustic practices attempt to rework the distribution of the heard and the felt. In this sense, I may speak of *compositioning* as what works at the making of an acoustic milieu in order to engage the spatial, tactile, and sonic materialities within a "region of objectivity." Working at the limits and affordances of a given region, compositioning captures processes of improvisation and the dynamics of shared play, the assembly and ensemble of agents and actants across human and nonhuman worlds, and that perform certain instrumentation, certain assemblages in support of the alliances or affinities central to ontological organizing. In short, an ontological expression gaining traction by way of an acoustic performativity. As such, compositioning is imagined as the composing we do *on the way*, in the rush of life, and that aims at expanding precisely the limits of a given place; that is adept at moving across both actual and ghosted territories, from the material structures and matters of space to the compendium of frequencies and reverberant geometries defining as well as enlivening environments. Compositioning is a vertiginous work.

I want to draw out notions of compositioning then, following Papadopoulos, in order to engage more fully the question of acoustic politics—acoustics as a politics in and of itself. If acoustics, as a performative paradigm, as a knowledge of the echoic and the vibrational, the rhythmic and the transmitted, is a defining feature of the social, and the constructs of place, questions of imperceptible politics may assist in capturing the potentialities found in acoustic milieus and their arrangements: the acts of compositioning that engender particular refusals and possibilities, drawing upon the ways in which the acoustic is always tensing the limits of the perceptible. We may think of these acoustic arrangements as a *counter-pointing* to existing structures and situations so as to enable often emergent forms of togetherness and affinity—the compositioning that often draws together a range of elements, from the near to the far, the actual to the virtual, that works through that multiplicity of trajectories of sonic experience: *a disordering that orders otherwise.* Compositioning may run across centers and margins, strangers and neighbors, to figure particular acoustic milieus through which experiences of belonging or unbelonging, the syncopations and disjunctions of place, may be creatively and critically engaged.

I'd like to delve further into notions of compositioning by considering the initiatives and expressions found within activistic situations, particularly within communities bound by a culture of collective making, of commoning. The occupied squat and social initiative La Tabacalera in the neighborhood of Lavapiés, Madrid, may act as an example, a site or scene allowing for a consideration of how compositioning often assists in self-organized movement, lending palpable input into the improvisational and deeply attentive caring such movements require. La Tabacalera formally developed in 2010 from out of the economic crisis spanning across Europe at the time, and in particular, in response to the housing crisis in Spain, which led a group of local activists and neighbors to negotiate the possibility of occupying the empty tobacco factory in the center of the city. Having been closed in 2000, and finally given over to the independent association (Lavapiés Cultural Association Centro Social Autogestionado La Tabacalera) in 2010 with the suspension of an architectural competition (due to the financial crisis), La Tabacalera has worked to set up and maintain a range of activities, including cultural and social workshops, and the hosting of meetings and debates in collaboration with various social and political organizations. This has led to solidifying La Tabacalera as an important social center and cultural meeting point for the city, and those active in local struggles and political movements. La Tabacalera continues to exist as a self-managed, independent social center consisting of workshops in silkscreening, bicycle building, urban gardening, music and theater practice, along with ongoing educational and neighborhood outreach activity.

Central to La Tabacalera is the bi-weekly open assembly, which is held in the main meeting hall in the building (named *Sala sin jefe*, or No Boss Room). The assemblies are open to all and are aimed at addressing current issues as well as proposals and ideas from the larger community. The assemblies come to articulate the center's embedded commitment to direct democratic forms,

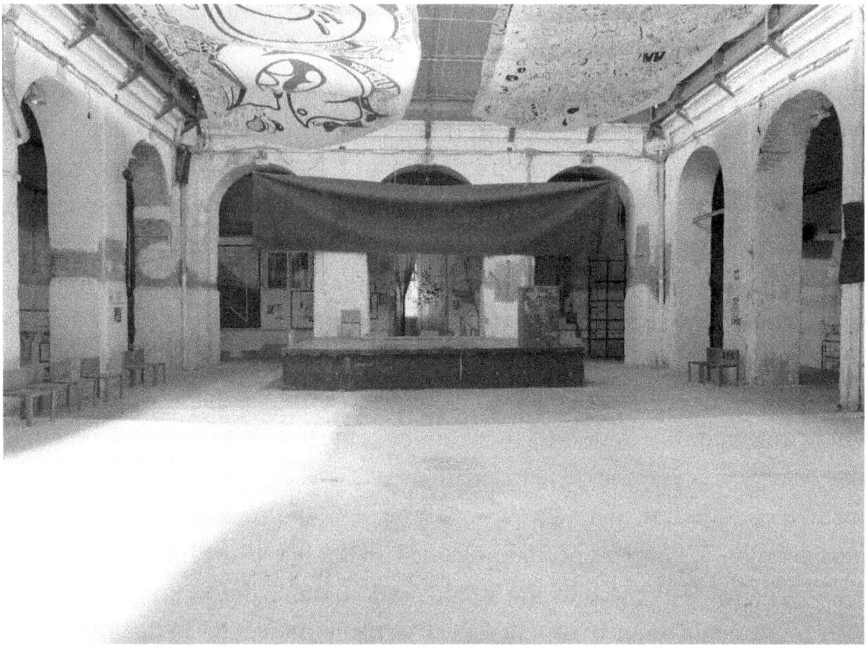

Figure 1.1 Assembly hall, La Tabacalera, Madrid, 2017. Photo: author.

horizontality, transparency, and co-responsibility, making the managing association secondary to the assemblies as a governing form. As such, La Tabacalera sets out to embody a living expression of participatory governance, which is reflected in its structural forms as well as in how people may contribute and work within its spaces. In this regard, the assembly hall becomes emblematic, articulating by way of its openness, a dedication to participatory governance. Such a position is brought to life in the very enactment of the assemblies, which must be appreciated on some level as an acoustic expression, the articulation of an acoustic arena that specifically facilitates the movements of voice and the hearing of views. Central to participatory governance is the crafting of such an acoustic dimension.

My own experience as a collaborating artist in 2017, working within the context of the gallery which operates in one-half of the building (run by the Ministry of Culture), was deeply shaped by observing the general ethos expressed by the community at La Tabacalera.[37] Throughout the process of developing an artistic project, I was interested to consider the ways in which questions of citizen participation and direct democracy come to be expressed in the environment of the occupied building. The arrangement of chairs in the front lobby of the space, the open assembly hall in which the activities of the center are debated and defined, the pin-up board presented at the front entrance giving notice of current activities, the various handmade signs and banners hung around, the graffiti and murals that fill the basement level, and the many workshops geared toward DIY knowledge and collective making, these aspects are woven together to articulate a practice of composition that moves from how bodies occupy and foster an environment of inclusion to the creative processes by which materials are worked through as questions of sharing and sustainability.

What I'm considering are the ways in which a general ethos and social sensibility are fostered not only through articulated statements or protocols but also through the arrangement of chairs and studios, signs and objects, meetings and workshops—the timings and spacings that are critically arranged and that support the potentialities and challenges at the heart of horizontality. These contribute to an imperceptible politics which, as Papadopoulos suggests, supports and extends more "frontal" forms of political engagement. The particular arrangements of material things, and the crafting of spaces, participate in a larger grammar of gestures and movements that become reflective of an attitude, a social sensibility geared toward the interdependencies that make up social bonds. From material constructs to gestures of sharing, the crafting of inclusive spaces and techniques, these become expressions of compositioning, which are equally aesthetic and functional, theoretical and technical, structural and improvised—and work to facilitate the project of commoning and the ongoing expression of a social body in the making, a "living process, a collective learning experience."[38]

In his research and writings, architect and urban activist Stavros Stavrides highlights a set of spatial scales around which commoning as a practice can be appreciated.[39] From the scale of territory, and the making of common lands, to housing, and the construction of shared living arrangements, and finally, to a question of "objects," Stavrides opens a productive understanding of commoning specifically through terms of scale. In particular, I'm interested in his understanding of objects and how objects figure within practices of commoning. According to Stavrides, while territory and housing are more obviously expressive of commoning practices, objects are also essential because objects often circulate, performing within daily routines and shared activities; they engage us on a bodily, intimate level, and may also require constant care or repurposing: from cleaning to arranging, passing to others or working with, modifying and rebuilding, objects are touched and shared, and as such, they capture and facilitate the general flow of life in the making. Practices of commoning also emerge through the arrangements we make of objects. Arrangements of objects impart onto the physical attributes of a space or territory a material, compositional process or decision-making, where a set of chairs placed in a circle, for example, articulates an approach to inhabiting an architecture or environment collectively,[40] or where open-source technologies facilitate the crafting of common tools. The arrangement and crafting of objects, in this regard, acts to nurture a sense for the sharing of property, and the work of horizontality, and is therefore often central to supporting cooperative values and actions.

Extending from Stavrides' thinking, the arrangement of objects and the shaping of a common milieu are important for facilitating the power of people, and the ways in which shared spaces are maintained. Objects and infrastructures, material formations and spatial constructions, all contribute to the practice of commoning, and additionally include the making of structures and situations by which one may speak and be heard by others. The public assemblies, the reverberant and open space of the meeting hall (which has *no boss*), the capacity to enter and access workshops, or to find ways of participating, all come to enable the general movements of speaking and listening, of feeling and being together, of experiencing an affective sense of recognition and resonance that work to foster and enrich an ethos of common affection and the making of life together.

It is my perspective that acoustic practices, and the sense of the distribution of the heard, contribute to the crafting and placement of objects and their performative circulation, and dramatically aid in the perceptible and imperceptible movements that support an environment of cooperation. From the vibrational sense of commonality, which impacts onto the material shape of things, affording a sense of graspability, to the echoic differentiation enabling individual argument, a space of sensitive debate and the cultivation of critical consciousness, practices of commoning articulate an acoustic performativity on many levels. This is equally figured if we consider the ways in which such collective and common initiatives are dependent upon an intense cultivation of attention: how processes of mutual care, participatory governance, and collaborative learning are built upon the powers of attention in which listening is extremely operative. These situations that, as Silvia Federici suggests, require a listening with the whole body.[41]

In relation to La Tabacalera, as an artist I was interested to materialize my project around the idea of acoustic commoning[42]—the ways in which acoustic practices give support to reorienting our political or social selves toward forms of collective making, participation, and horizontality. In this sense, I felt moved to amplify what I recognized as an essential aspect within the environment of La Tabacalera, that of the acoustic dimension or performativity directed toward commoning. Here, I may speak of *acousticking* as a process of "acoustic crafting"—the shaping and reshaping of spaces and durations as a process of caring for what is needed in the pursuit of collective power; an acoustic craft adept at the skillful nurturing of care and solidarity so essential to the tender project of common space. While spaces of collective organization, as resistant movements of togetherness, are necessarily strongholds against the general capture defining neoliberal capitalism, they perform such resistances equally through pronounced forms of mutual care and cooperative labor. Opening up and maintaining shared space is to attend to the affective economies of attention, intimacy, shared passions, as well as that of uneven power, exhaustion, even collapse. Acoustic commoning can be appreciated as lending to these greater labors and tender movements, and is expressed through what I understand as gestures of sonic warfare as well as acoustic welfare: from the noises that work to punctuate and combat the forces that capture and evict to the generosity and sensitivity performed in support of attending to each other and which work at keeping sound and listening open as social resources.

During my initial visit to La Tabacalera I was shown around by an administrator from the Ministry of Culture, along with a representative from the self-managed community; investigating the workshop areas, informally meeting coordinators, and hearing about the history of the social center, I became curious as to the two upper floors of the building, which we did not enter at the time. In realizing that these upper floors remained unused, and were generally off-limits, I decided to focus on the issues of commoning, self-organization, and horizontality by proposing a creative occupation of the upper floors (spanning around 15,000 square meters). My idea was to convene a group of collaborators to undertake a performative experiment, a creative lab focused on collective movement, or movement strategies aimed at commoning. To do so, I worked with a group of local dancers, actors, and performers to participate in workshopping what it means to nurture a collectively determined space, and, further, what it means to occupy an abandoned building in the city, to engage with what AbdouMaliq Simone calls "the uninhabitable." The notion of the uninhabitable, for Simone, allows for drawing out the complex knot defining "improvised lives" existing within and through precarious environments: environments of dereliction, of abandonment, as well as environments stressed by the geopolitical force of capitalism, which impact particularly on the Global South. For Simone, the uninhabitable is tensed between economic capture and daily improvisations, poised as zones of abandonment that are lived in nonetheless, and that give way to what he calls "thousands of small experiments." These experiments work at reconciling states of precarity through the very conditions of precarity, and in doing so, they twist together an art of survival, or "forms of fugitive graces."[43]

Over the course of a week, we set out as occupiers of the upper floors, confronting spaces seemingly uninhabitable—even gaining access required permission, and was greatly restricted due to potential hazards. Within the context of the artistic work, questions of habitation and the uninhabitable, abandonment and occupation were set in motion as both material, concrete realities as well as fictional and speculative processes. This approach allowed for a flow of activities to unfold, from entering the building and accessing various rooms, rummaging through the trash and left-behind detritus scattered throughout, to cleaning particular areas, collecting together potentially "useful" materials, and creating different expressive actions together—allowing for a breaking down as well as a building up. These activities were enacted as performative gestures, as improvisations, where materials and objects became partners in "thousands of small experiments": mobilizing collective movements,

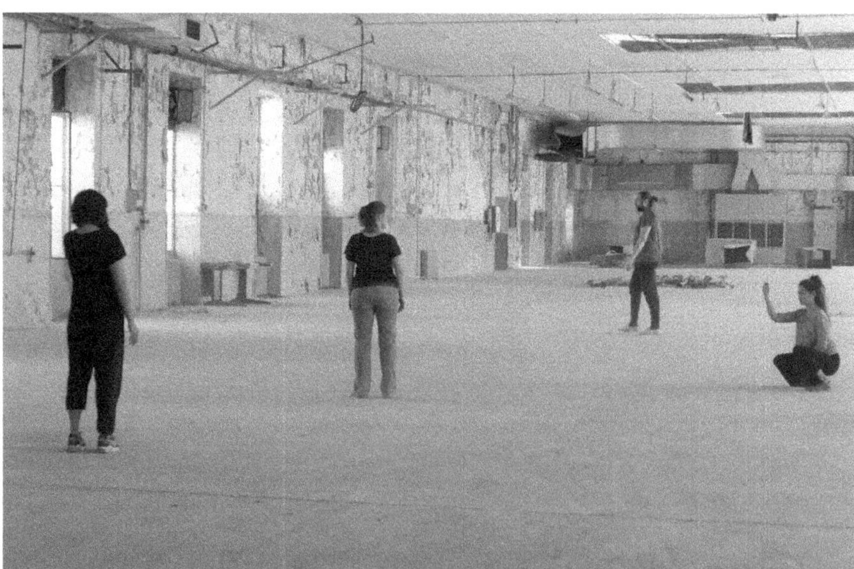

Figure 1.2 Collective movement lab, with Vicente Colomar, África Clúa Nieto, Annie Pui Ling Lok, Catalina Mahecha. La Tabacalera, Madrid, 2017. Photo: author.

in touching and breathing the vast open spaces, including all the dust and textured surfaces, building up strange material constructions from found wood and metal, rope and cardboard, disused and broken things, that became shelters, a garden, a weird form, or a scenographic backdrop to choreographic experiments and digressions, a losing of oneself and a gaining of something else.

My interest throughout was to pose the question of occupation and horizontality as foremost bodily actions, a physicality and sensuality, an orientation guided by material crafting and embodied knowing that aimed at figuring an emergent sense for being in space, of being in touch with space through each other. In short, to reform a sense for space as a collaborative matter and process of invention.

From this left-behind matter
To the sweaty heat that pulls at the breath
Breathing in while moving out
Steps punctuating time, the time of loneliness
This loneliness shared suddenly
Or maybe it is only an illusion
What can be pulled from this left-behind matter
Into constructs of generosity, is that right? I wonder . . .
With the emptiness becoming full, full of not knowing
The unknowing that seems so necessary
To move and to hesitate at the same time
As the hand digs into the trash, tracing another's back, pulling at his shirt
Running to some wild rhythm.

The week unfolded as a series of shared movements, a *compositioning*, a counter-pointing to the rhythms of the city outside, touching materials, working or, rather, *unworking* together, improvising, imagining what this space could become or what it might allow—how it might give way to social formations defined collaboratively—as well as questioning and discussing what occupation might mean in the context of Madrid and elsewhere. This included sharing and discussing how the local participating artists were also surviving in the context of austerity; how many of them were involved in cooperative projects and communities, local actions as well as political parties (such as Podemos) and citizen mobilizations (for instance, around Ada Colau and the movements in Barcelona).[44] Such concrete realities found their way into our movements in and around the abandoned floors of the disused factory. And in reflecting on how the desire for other ways of living, of commoning as the making of shared spaces

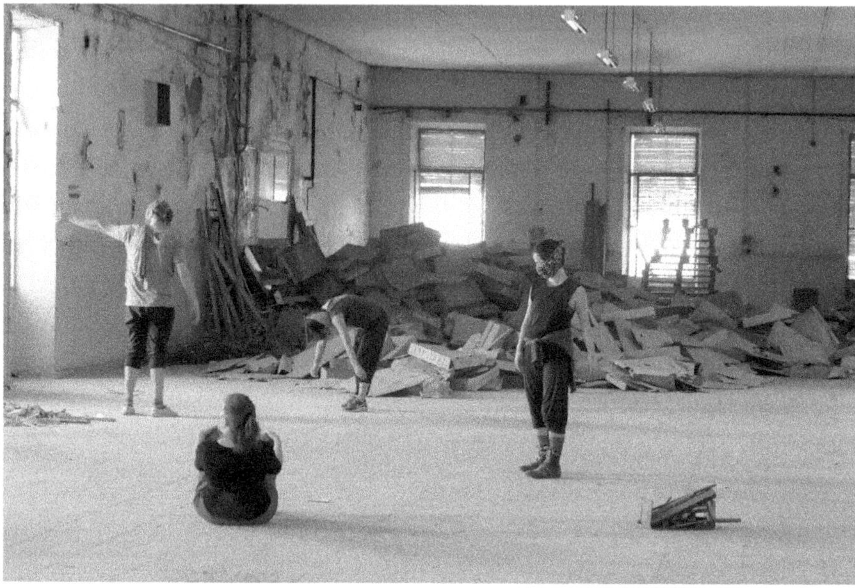

Figure 1.3 Collective movement lab, Vicente Colomar, África Clúa Nieto, María Escobar, Annie Pui Ling Lok, La Tabacalera, Madrid, 2017. Photo: author.

and situations, necessitates particular organizational structures and positions. It quickly became clear that spatial decisions are equally ethical questions, about how to make and give space, to be in space over time together, especially spaces that are off-limits or precarious, and to enable another's movements into and through an environment. Such bodily, spatial, material, and ethical experiments lent to a sense for the modeling of a community to come. As such, there was always this sense of risk, a type of danger, as to what might emerge by going further, into the uninhabitable maze of dark corners, trash-filled corridors, with crashed-in ceilings and left-behind matters: by staying close to community as that which must remain partially unnamed. As Giorgio Agamben suggests, "A true community can only be a community that is not presupposed."[45]

Throughout the course of the workshop, issues of sounding and listening, of acousticking, were discussed as well as performed, leading to a sensitivity for voice as well as noise. Often speech would turn toward other vocalities and other modes of conduct—a sounding of the space through shouts and screams, humming and melodic refrains, that often passed from one performer to another, expanding and then contracting. Voice and vocality were often about supporting each other, animating the larger collective movement as a transformative endeavor—this voice that celebrates this moment, that billows out to accentuate this emergent sociality, this improvisation. From vast open floors to darkened corridors and interior rooms, the acoustics of the building was often punctuated by voice and the noise of bodily actions: stomping the floors, scraping the walls, drumming found objects, throwing things or building types of instruments to perform as part of a larger choreography or ensemble. At the same time, there were moments when we collectively listened, attuning to the sounds from outside as they reverberated across the space and ourselves: the sounds of cars and the noise of daily life flowing by.

These gestures of sounding and listening often gravitated toward a form of "musicality"—a compositioning that became essential to the collective gathering, capturing, and performing that sense of attunement and orientation listening comes to support. Experiences and expressions of rhythm and echo, of synchronizations and differentiations—of joining together as well as breaking off—these emerged as gestures defining the sense of shared participation. Acoustics, in this situation, both grounded us within the space, enabling a feeling of togetherness, a tangible materiality held *between*, a punctuation of place (even a celebration), while affording a sense for deepening relations to the surrounding environment—a movement, a crafting, a reverberant flow passing between here and there, the near and the far, the actual and the virtual. Compositioning thus worked at commoning by fostering an emergent sense for sharing and building something together, a sticky endeavor that necessarily relied upon a certain dizziness or disequilibrium by which to imagine otherwise. Acousticking, in this regard, may act as an energetic support for reorienting the conditions defining and shaping forms of life, allowing for other spatial and temporal geometries, other ghosted figurations to appear—"strange alliances" as Simone says, that support other modes of inhabitation.

Returning to Papadopoulos, practices of composition work materially as well as ontologically, and can be understood to reorient the conditions of particular social and structural formations: to interrupt the ways in which bodies are figured by a given patterning—the spatialities and temporalities, the availability of resources and institutional access, the communal flows and movements by which one may experience a sense of belonging or unbelonging, possibility and impossibility. Compositioning, as I'm suggesting, draws from such conceptualizations while also turning us toward the acoustic as a particular ontological base. Compositioning is a critical and creative endeavor aimed at capturing what may be at hand, explicitly retuning or detuning certain acoustic

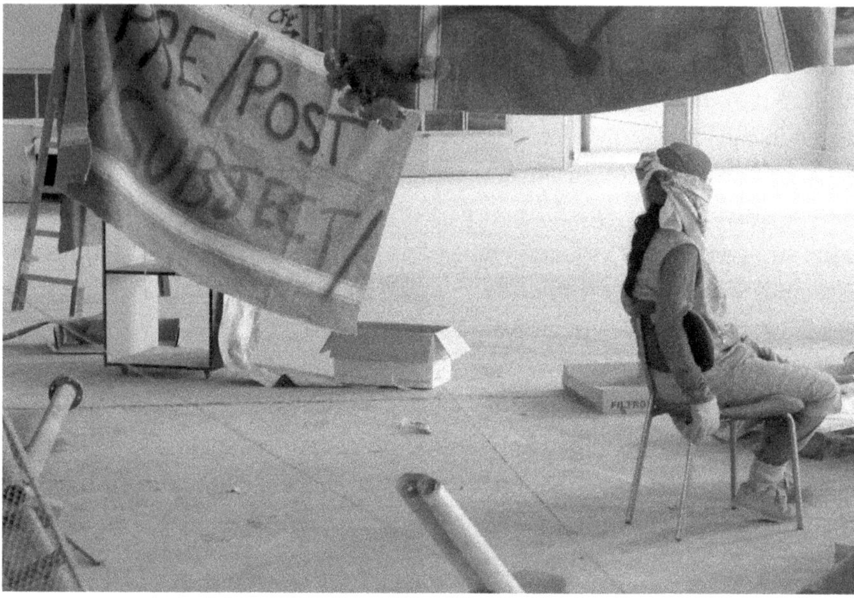

Figure 1.4 Collective movement lab, Fátima Cué, La Tabacalera, Madrid, 2017. Photo: author.

norms and sonic matters, making them available for oneself and others through improvisations and sticky arrangements, and refiguring the systemic tonalities that often condition the movements of bodies and communities: an acoustic welfare in support of common life.

In working in the upper floors at La Tabacalera, during that one week where a small group of people undertook a type of rehearsal—a performative playback and remix of what was going on downstairs, within the self-managed squat community—there emerged a greater sense for the significant interplay between space and collectivity, voice and feeling underpinning an ethics of horizontality and the work of commoning. As with the community at La Tabacalera, sound and listening may operate as situational processes for reworking relations between the near and the far, the actual and the virtual, matter and its transformation, between how one speaks and how one feels the speech of oneself as part of a greater community effort. Acoustics comes forward as a performative base, a crafting which gives life to the fundamental struggles around space, the time of shared labor and togetherness, and the movements of social solidarity that rely upon a constellation of matters. Here, the reverberant assembly hall at La Tabacalera sets the conditions, real and figurative, for the flourishing of a participatory citizenry, a worlding endeavor that gains traction and support from the shared volume of a collective voice—an energetic feedbacking by which ideas and their sharing gain intensity. From the fundamentally pragmatic and local to the articulation of larger governing principles, facilitating and sharing in discussions is crucial and key. Such acoustic commoning is often the socio-material basis from which one acquires a sense for the possibilities of what we may compose, from sonic warfare to acoustic welfare. For there are also often moments of exhaustion, of deflation, and frustration. During the week of activity on the upper floors, things were not always so easy as the group rummaged through the dusty trash, sweating in the heat gathering under the Madrid sun. It was an exhausting, vertiginous week, a week of not knowing for sure what we were necessarily after, what "the project" was about—we worked at it, and it worked us. Within the activities, we also needed to take care, to take time out for resting and for talking about other things; commoning is also a tiring work, an unworking of the usual flow of how we spend time. As Carla Bergman and Nick Montgomery thoughtfully remind, horizontality is something we must learn, which includes learning to set limits for oneself: one cannot *give all* all the time.[46] Rather, a mode of pacing is needed, a sensitivity for nurturing weapons of combat as well as those of care, a carefulness as to the energies given and taken, for crafting matters of support. Listening to oneself and others is often about such care-taking.

Acoustic Justice: Commoning and Violence

As I've set out to argue, acoustics may be positioned as a critical and creative framework for understanding how one navigates the conditions of particular places, and how one may seek out and construct paths of (re)orientation, which is often related to struggles over recognition, belonging, and the figuring of oneself and others. In this regard, acoustic performativity is a process as well as a conflict over space, as something that shapes us, contributing to who may enter and what types of behavior find welcome. Place-making is fundamentally a contested and complex process, aiming to capture a given space as a performative zone for playing out the drive and needs of communities; whether in the occupation of abandoned factories, the self-organized spaces of home, or in the collective making of clubs, space is fundamentally enabling for acts of joining together and for living according to what one holds dearly. The acoustic shape and performativity of such spatial acts often work to support the movements of participation and the coexistence of all that is gathered together, emboldening the material figuring and potentiality of common determination. Hearing and being heard, attuning to particular voices and movements, echoing and synchronizing around certain social imaginaries and rhythms, come to impress upon the contours of space a power of listening—an attentiveness for what might emerge from out of the reverberant geometries animating the spaces around us.

The issue of participation, or what Nancy Fraser terms "parity of participation," may be considered here.[47] For Fraser, it becomes imperative within a framework of justice to balance understandings of social and cultural recognition along with the issue of economic disadvantage, that is, to draw out a critical theory of justice that would recognize the parallel tracks of recognition and redistribution. On one hand, this includes appreciating social recognition as a process of making distinctions, of capturing recognition by way of difference, arguing for cultural specificity: an identity politics made in claiming recognition. At the same time, questions of redistribution require a concern for social equality as an economic issue. Redistribution, within the context of justice, draws focus to the socioeconomic, and addresses the necessity to grant opportunities by way of economic access and resource across society, regardless of difference. For Fraser, these parallel, and seemingly incompatible, views of justice demand a more pronounced sense for building alliances, for instituting structures aimed at contending with the injustices of social misrecognition and economic disadvantage.

Returning to La Tabacalera, and the question of participatory governance and the occupation of space, it may be seen how expressions of recognition and redistribution are brought together around the project of commoning. How the crafting of common space negotiates the particular struggles over housing, and the crisis of economy, creating other conditions of economic access and distribution through forms of mutual care and cooperation. Such commoning focuses on a "solidarity economy" model, one which equally articulates culture as a collective resource: not something to be distinguished through an insistence on difference, but rather, one that specifically fights for "free culture"—an open-source culture that works at nurturing collaboration. Redistribution and recognition are thus brought together through commoning practices and the instituting of a common world.

Compositioning, and the material and ontological organizing Papadopoulos sees as underpinning practices of composition, may be underscored here, to appreciate how communities may work at a "parity of participation" as Fraser calls it: at formulating a material and social construct that may help in contending with marginalization and inequality.

Participatory governance seems to figure an expanded activity, composing by way of collective space, the crafting of materials and tools, organizational acts and principles in support of horizontality. These are woven together in building alliances, and which suggest an activity comprised of "thousands of small experiments" that interweave what is near, what can be found, with the distant and the virtual: an open-source framework always prone to cooperative modification, repurposing, improvisation. Here, what an acoustic performativity may point to is how sound and listening, as capacities for attunement and interruption, consonance and dissonance, greatly enrich the work of participation in a variety of ways and, therefore, may be understood as resources.

In addition to commoning, and the nurturing of shared worlds, I want to conclude by considering how acoustics contributes to forms of forensic analysis and crime reconstruction in order to address questions and realities of injustice. Acoustics, in this sense, is applied to the task of constituting scenes that are often hidden from view, or that contain only the left-over debris of particular events. The process of acoustic modeling or reconstruction may allow for probing situations not only of communal togetherness but also of terror and the ugly work of torture. As such, it opens another important view onto acoustic performativity: within such troubling scenes of violence, acoustics may assist in penetrating the dim and dark spaces in which state and humanitarian crimes are often enacted, and that continue undercover and beyond international view.

Forensic Architecture, a research collective based at Goldsmiths, University of London, has worked extensively at addressing situations and scenes of human rights' violations and state violence. Their work is often based on developing methods of spatial and visual reconstruction and modeling, drawing together visual, material, and auditory information in order to trace the causes of certain events. In particular, their work on the Saydnaya prison in Syria has focused on the use of acoustic material and sonic information in constructing an understanding of the prison and the human rights' violations taking place there.[48]

Saydnaya prison is located north of Damascus in Syria, and since the beginnings of the uprisings in 2011, and subsequent civil war, has operated as a central detention and torture center of the Assad regime. The prison is inaccessible to journalists and the larger international community, with the government of Syria obstructing any attempts at gaining information. Rather, the prison remains off-limits and exists as a black hole within the larger view of the state's activities and conduct. In collaboration with Amnesty International, Forensic Architecture has worked at building a website depicting the architectural and spatial layout of the prison. Developed primarily through testimonies from former prisoners, the group has rendered the prison through a series of animated digital models, which includes an extremely poignant application of sound and acoustic information (in collaboration with the artist Lawrence Abu Hamdan).[49] As the former prisoners recount, experiences of being in Saydnaya were dramatically shaped by the strict control of sound and silence within the prison. The interior of the prison is kept dark, with prisoners held in darkened rooms, either confined together in large groups or in solitary cells. Unable to see clearly, the prisoners are forced to keep silent. Even the smallest whisper or breath, even a cough, could lead to beatings by the guards, and even death. In addition, when beatings do occur prisoners are not allowed to make a sound: "In other prisons, the guards wouldn't leave the prisoner alone until he screams. If the prisoner doesn't scream the guard would take it as though the prisoner is challenging him. He has to yell. Saydnaya is completely opposite. If you yell, the beatings would intensify. If you keep quiet it would go down until he finally stops."[50]

As a consequence, such an enforced condition of silence produces an extreme level of stress, marking every sound, every breath with anxiety. Held in the dark, prisoners are made to fear what might emerge from out of the shadows. Within such conditions, sound and silence are terrifying. Whether through the sounds of prisoners being beaten, the amplification of pro-Assad songs blaring throughout the reverberant corridors, or the deeply tensed silences prisoners fear to break with their own cries or whispers, the acoustic life of the prison is geared toward maintaining a state of violence. This extends through the very fact of prisoners needing to keep silent. Abu Hamdan makes the observation that, in keeping silent, prisoners must also restrict their own movements; in this way, one is not only guarded by an external presence but one also guards oneself.[51]

Forensic Architecture's project captures the larger acoustic as well as architectural dimensions of the prison, reconstituting, through computer modeling and animation, the experience of being inside Saydnaya. This both serves the purpose of showing the prison, unmasking its hidden activities to a larger international community, while also recounting the stories of those who have suffered behind its walls. Acoustics is utilized here so as to penetrate the darkness and disarm the obfuscation of accountability.

The artist Lawrence Abu Hamdan, a participant within the Forensic Architecture project and collective, has persistently worked at issues of audibility and inaudibility, and the ways in which acoustic information, digital eavesdropping, and voice analysis are being utilized within criminal investigations and judicial procedures. In particular, Abu Hamdan is concerned to bring into view questions of state violence and how a politics of listening is brought to bear upon justice work. This has led Abu Hamdan into a range of compelling research and projects. From questioning the emergence of new forms of "expert listening" within the field of vocal and accent analysis (used to determine the origins of witnesses or migrant detainees), to acoustic reconstruction as found within courtroom proceedings that rely upon understandings of sonic evidence (for instance, in the case of Oscar Pistorius, whose trial relied upon understanding how the sounds heard from behind a wall led to the killing of his wife)—such research and issues are captured to draw out complex perspectives onto what constitutes sound and how aurality impacts the understanding of evidence and crime. Abu Hamdan's "aural forensics" is further applied to the Saydnaya prison. As a process of translating prisoner testimonies into acoustic modeling, and attending to the listening experiences of prisoners as evidence, Abu Hamdan shows the ways in which sound is put to use within scenes of state violence.

In addition to the Forensic Architecture project, Abu Hamdan has developed a set of related works unpacking the troubling question of the Saydnaya prison. In particular, he has produced a set of "ray tracing" drawings depicting the propagation of sound through the prison architecture. Ray tracing is a technique used to show the distribution and reflection of sound under given architectural parameters; in particular, it captures "sound leaks" which, in the context of the prison, begin to outline the experience of hearing from within the cell. The acoustic exposure performed by the drawings begins to render the prison as a construction of walls, barriers, and spatial limits, and additionally, as a set of flows and intrusions, acoustic reflections and their immersive potential. In reading the drawings, we are led to understand in what ways sound is utilized in order to traumatize prisoners, to hold them within a violent state of fear through acoustic means—the violence announced and enacted in decibels, and the unseen yet deeply registered sounds of torture and murder.

Under such rendering, incarceration takes on another look, one that allows us to gauge the experiences of torture as an act of sensory violence. Here, the acoustic and reverberant geometries discussed earlier, as a ghosting of architectural space, are turned toward fostering an ecology of fear in which confusing the location of sound sources serves to traumatize—or where the multiplicity of trajectories articulated through sonic spatiality are channeled and controlled in order to harm. At the same time, as Abu Hamdan notes, such acoustic intensities forced prisoners to detect their surroundings by way of sound, thereby heightening their level of listening. As earwitnesses, the prisoners are able to give testimony to many aspects of the prison and its activities, allowing for greater understanding as to the crimes being perpetrated. "Silence was used to accentuate the prisoners' oppression and restricted their abilities to speak and circulate knowledge, yet paradoxically, it produced heightened modes of listening that led to the discovery of crucial information about life in Saydnaya."[52] Earwitnessing becomes an important position and capacity within an acoustic justice framework, underscoring not only the emancipatory potential in crafting acoustic commons, and states of participatory governance, but equally in drawing forth evidence within scenes of violence.

As I've tried to argue, acoustics may be captured as the distribution of the heard, in which the movements of sound and the related capacities of individuals and communities to form bonds, or enrich collective identities, are worked through. From vibrational sharing to echoing differentiation, rhythmic synchronizations and transmitted projections, acoustic practices come to effect articulations of orientation, belonging, and escape. These become compositional efforts, whereby forms of sociality are fostered or unworked in order to support struggles over recognition, a tuning and retuning of the particular tonalities or alignments that support oneself and others. Compositioning is therefore often a form of acoustic commoning, an acoustic crafting in which sound and listening support the formation of shared worlds: the solidarities and alliances captured by an acoustic performativity, which is often defining the experiences of hearing and being heard so essential to being-in-common.

In following the projects on Saydnaya by Forensic Architecture and Lawrence Abu Hamdan, it becomes more clear how acoustics impacts onto one's sense of orientation, bodily and mental support, and securing a sense of place. Yet, in this case, such capacities are put to use for an extremely different purpose. Acoustics, as the situation in Syria demonstrates, can be deployed as an effective instrument of abuse and violence: to capture and control through enforced silence, which comes to weaponize the reverberant geometries of propagating sound, and by way of the rhythms that intensify an overall ambience of fear with a sense for pain to come. Here, fear and anxiety are palpably materialized by the intensely affective, pervasive, and traumatic force of sound and its acoustic deployment. Acoustic performativity thus goes both ways: from the practices of commoning and the nurturing of sensitivity for others to the violence aimed at harming people. Acoustics is revealed as dramatically shaping and contouring the experiences of being a body in space, orienting one through particular acoustic acts and designs. Or, as Abu Hamdan also articulates, it may be utilized in the gathering and articulation of evidence, allowing for critical access to the otherwise hidden scenes of violence.

The practices and performativities that enable or capture hearing and being heard are thus held within a larger question of acoustic justice. Acoustic justice is about the conditions and economies that shape aurality, and the degree to which acoustic arenas come to restrict or support social and political possibilities; it is about the capacity to articulate life stories and the expressive orientations of diverse manners of living; to feel oneself as part of the continual work justice entails: the work performed to support conditions of equality, of mutuality—to cultivate just relations. Acoustic justice is also aimed at reorienting attention toward situations where life is under threat or unjustly captured. Being heard is often the basis by which to build self-esteem and processes of self-realization; it is also what allows for being a witness to others, to attend to the accounts of others, within a range of contexts. Acoustic justice is therefore deeply shaped by a politics of listening, and by extension, the acoustic arenas or norms that support or deny recognition and participation. Whether in establishing particular cooperative forms of life within abandoned buildings, and in support of an ethics of commoning, or in working to constitute or reconstruct hidden scenes and architectures of human rights' violations and state violence, acoustic justice works at hearing what is often underheard, bringing this into a time and space of concern and criticality, a scene of telling and answerability. It draws out struggles over recognition and accountability that may also allow for processes of repair—of acknowledging others as well as nurturing what we may still create.

The former prisoners of Saydnaya, in guiding us through the darkened interiors of the prison, ask of us to hear what they heard. Such concerted attention aims at disturbing the given order of hearing kept in place by the Syrian government: by what we are allowed to hear and to know. What consequences might follow from such hearing, such attunement, and the larger project of acoustic justice? As Lawrence Abu Hamdan suggests, addressing sound and aural information as evidence supplies us with a tool for penetrating certain darknesses, particularly those in service of state violence, allowing for other ways of getting at the complex and often cruel mechanisms at work against struggles over rights and recognition. In hearing what others hear, we also extend what counts as the common world.

2

Poetic Ecologies

Resonance, Imagination, Repair

The social and political dimensions inherent to acoustics find articulation in a range of gestures, enactments, and practices. From the ways in which speech and listening contend with the acoustic norms setting pace to audibility, to the expressions of power garnered by way of volume and its situated force, acoustics captures possibilities and challenges surrounding struggles over recognition and belonging. Such possibilities and challenges underpin the concept of acoustic justice.

In questioning acoustics in this way, I also work in and around notions of relationality—whether in considering the sense of agency and empowerment found in forms of social commoning, or in the capacity to extend the reach of listening toward the nonhuman, questions of relationality are fundamental. In this chapter, I'm interested in bringing into critical play a number of perspectives on relationality, underscoring *acoustic relationality* as a particular framework. What pathways of relationality are made possible by way of acoustics? If acoustics is understood as that which facilitates or fosters the movement of sound, and the crafting of a sonic image, around and by which acoustic communities find traction, what particular understandings or conceptualizations of being-in-relation might emerge? And further, what relational capacities or imaginaries are afforded by underscoring listening as multimodal, as facilitating forms of care, compassion, and understanding that extend beyond the strictly aural?

To engage such questioning, practices and works of sound art may act as a focal point, allowing for a consideration of the ways in which acoustic relationality is made manifest. Throughout this chapter, sound art is reflected upon as an arena of sonic aesthetics, the crafting of sound as a contextual material, as well as a broad field of expressions that figure experimental and engaged acts of relational work. In this sense, I'm aiming to bring attention to the ways in which acoustics, as the distribution of the heard, is mobilized through works of sound art in order to capture or foster relational intensities and potentialities. In activating experiences of sounding and listening, and experiences of attunement and interruption, works of sound art also extend relationality, and understandings of recognition, beyond that of the human, widening the sphere of what counts as legible and intelligible. Subsequently, methods and expressions of sound art reveal a tendency for approaching the foreign, the strange, and the unfamiliar, which often leads to an experimental engagement with architecture, place, voice, the nonhuman, and the world of matter. In this regard, sound art, and the relational trajectories it comes to manifest, elaborates acoustic justice by contending with the limits of social equality in particular ways and contexts.

Poetic Ecologies of Resonance

Historically, I may turn to Alvin Lucier's work *I Am Sitting in a Room* (1969) as a way to open the question of acoustic relationality. *I Am Sitting in a Room* is based on the composer recording himself speaking, narrating a statement, then playing this back into the same room while re-recording, again and again, for a duration of thirty-two cycles.[1] In doing so, what comes forward is a slow dissolution of the composer's articulated speech as the room's acoustic resonances become more pronounced, until at the end all that remains is an undulating sonic drone. Lucier's speech is altered through this acoustic operation, where the resonant properties of a room capture or extend his voice, *overtaking* the articulation of speech.

Lucier's work (along with many of his compositions and installations throughout his career) captures an aspect of acoustics that I would like to consider, that is, the inherent instability found in the trajectory of sound as it moves through space. The performative undertaking central to *I Am Sitting in a Room* captures sound as site-specific and acoustically situated while having the capacity to untether itself—to alter *along the way*. Although Lucier relies upon an electroacoustic, technical apparatus, the act of recording his voice and repeatedly playing it back underscores a certain propensity of sound and, by extension, acoustic behavior in general: sound as a matter of propagation.[2] This *extending away*, this itinerancy, this fleeing from a source that sound performs, and which partly defines sound as such, is evocatively captured as the basis for *I Am Sitting in a Room*.

There exists a potentiality found within sound's propagating movement: a type of flexibility, a connectivity, even an erotic charge from out of which other exchanges and manifestations may take shape. From the source of sound to its diffusion into the environment, one may identify a potentiality of emergence—this propagating movement that incites movement itself, a molecular and material agitation, giving way to sympathetic, resonant, or reflective responsiveness. Acoustics, accordingly, may be emphasized as the conditions or constructs that enable the movement of sound through an environment, and in doing so, lends to this potentiality: for what may emerge from out of its reverberations.

In this sense, between a source of sound and its propagation a potential for transformation is captured: an animation or flourishing that gestures toward *mutation: this sound that changes as it moves, that evokes a logic of transformation.* As can be heard in *I Am Sitting in a Room*, Lucier engages sound as a deeply flexible material whose acoustic dynamics offer a number of possibilities: from questioning the nature or constitution of architectural space, testing the limits of spatial definition, to experimenting with the tensions between sound and language, resonance and meaning, phenomenological fact and music composition, Lucier captures acoustics as the basis for a transformative process that subsumes "dualities" within a continuum of sound.[3]

Following Lucier's work, I'm led to consider how sound actively generates connections to the immediate and the situated that equally foster or evoke more ephemeral, temporal, and phantasmic relations to what we perceive. From such a view, sound comes forward as both sited *and* siteless. By bringing attention to the immediate or present situation, punctuating a particular event or body with its force, sound simultaneously diffuses into the environment. This rather ambiguous and generative dynamic—of site and sitelessness, of hard punctuation and soft dissolution—may be emphasized as inherent to the shaping of acoustic relationality.

In reflecting upon this aspect, I want to move further, to consider what may be drawn from this complex dynamic: What questions or discussions does acoustic relationality enable? How has the transformative logic of sound and sonic experience influenced the developments of sound art? And what further potentialities may be captured by way of the mutational chain of sound? Such questioning may be partly found already in the discussions and works occurring in the 1950s, in which the material emphasis found in the music of John Cage, for example, finds contrast with the more de-contextualized, non-referential electroacoustic works of musique concrète defined by Pierre Schaeffer, Pierre

Henry, and others working in Paris.[4] The relation between material objects and recorded constructs, between phenomenological fact and sonic (or electronic) transformation, in short, between the acoustic and the acousmatic, is mobilized as a central thread around which the emergence of sound art may be heard to take shape—that is, tactics and practices that spirit passage between material and sonic objects, field recordings and electroacoustic manipulation, between prepared pianos and stretched recordings, and finally, between the seen (acoustic) and the unseen (acousmatic).

Other composers and musicians, such as Pauline Oliveros, Daphne Oram, and Éliane Radigue in particular, equally developed a range of experimental and electronic music compositions and methods in the 1950s and 1960s that often worked across acoustic, recorded, and electronic sound. For instance, Daphne Oram, working in the context of the BBC Radiophonic Workshop, constructed an array of innovative electronic machines that enable abstract sound production by way of drawing directly onto 35-mm film stock, as well as incorporating both acoustic instruments and electronic sound within live presentations. As she writes in the opening paragraphs of *An Individual Note*: "We shall, therefore, be exploring aspects of electronics and acoustics in relationship not only to the composition of music, but also to the world in which we find ourselves."[5] Already there is a tendency, an ambition, an understanding of sound as relational, as what may traffic across acoustics and electronics, bodies and machines, supporting contact with the world. As Lucier further captures, sound moves, and in moving draws us toward what is *out there*. Such an approach—which Oram marvelously describes as "a-wandering, a-wondering, a-musing"—marks an important contribution to the debates between electronic and experimental music, and can be appreciated as setting sound art on a course that wanders and wonders between acoustic, recorded, and electronic processes and worlds—or more importantly, that embraces sound and listening as transformative.

Following the question of acoustic relationality, the work of contemporary artist Zorka Wollny is suggestive for capturing the acoustic dynamics I'm considering. In particular, her *Resonance Assembly—Composition for a Factory* (2014) initiates an acoustic procedure in which a group of participating musicians move through an abandoned factory located in Berlin while producing sounds with musical instruments, found objects, and materials, as well as with their own voices and bodies. Through the work, the group comes to form an ambulatory compositional collective whose sonic gestures act to acoustically enliven the disused space—or, as the artist suggestively states, to "bestow an acoustic presence to the factory's quiet erosion and untold stories."[6] Wollny's *Resonance Assembly* captures the possibility of reworking the contours and delineations of space, and the materiality of things around, by utilizing sound and acoustics, giving way to the creation of another sense of dimension: to conjure, to evoke, to assemble that which remains embedded within the disused site. As in Lucier's work, resonance becomes an acoustic means for engaging with the specifics of place, one that seems able at oscillating between situated understanding and musical rapture. Through such performative and acoustic collaborations, a type of generative disorientation arises, a destabilization of time and space, of the immediate, that draws out a transformative process or procedure.

Following Lucier's and Wollny's works, I want to reflect upon this acoustic articulation of site and sitelessness, this continuum of resonant sound that diffuses dualities, and how works of sound art draw from this complex dynamic, are fundamentally *driven by it*. From such works, acoustics is brought into a performative elaboration, forming an experimental base by which to expand upon that particular dimension/less quality sound engenders, and which situates us as listeners within a complex and somewhat paradoxical, vertiginous frame. For one is located within a construct that holds together site and sound, architecture and the spectral, and which activates listening precisely as a means for navigating this complexity. As such, the acoustic relation of sound and space, as the scholar Zeynep Bulut suggests, is one that dramatically articulates "the possibilities of movement, affect and change."[7]

Works of sound art, as I'm suggesting, often utilize site-specific methods, and a concrete approach to the textured matters of surrounding environments, in order to draw out another form of intelligibility, one that sutures materiality to the spectral or fleeting qualities of sound. In this regard, the experimental acoustics performed by sound art reworks conditions of space, bringing one into what I may term "poetic ecologies of resonance" in which the animate and inanimate no longer hold a steady, dichotomous position. Rather, spatial definitions and the specific demarcations of places are resounded, giving way to other entry points. In doing so, works of sound art often search for the foreign, the hidden, the untold, and the strange found within the architectures and environments around us, emphasizing the material facticity of things as being related not only to the apparent and the legible, but equally to the energies of invisible things, the alien matters of distant and even unseen locations, and the spectrality of the remembered, the not-yet or the disappeared. Sound, in other words, is captured as that medium enabling an engagement with what otherwise remains inaccessible, untold.

Poetic ecologies of resonance are suggestive for capturing this unique relationality—this acoustic potentiality that generates and gives way to the formation of other architectural experiences, other bodily orientations, as well as other pathways toward cohabitation, a *being alongside and with the unfamiliar, even the unknowable*. Resonance is posed, as Lucier and Wollny highlight, as the animation of static form, a performative transformation of space leading toward conditions of "musicality"—space as something *to be heard*.

In his philosophical meditation on listening, Jean-Luc Nancy offers an insightful view onto the topic of resonance. For Nancy, resonance, and the experiences of listening in general, acts as the basis for the emergence of "the subject," this body that, through the resonances that pass between inside and outside, moves across the matters of the body and the world, and comes to "summon" presence to itself.[8] As he outlines: "To listen is to enter that spatiality by which, *at the same time*, I am penetrated, for it opens up in me as well as around me, and from me as well as toward me: it opens me inside me as well as outside . . ." As such, the penetrative, expansive qualities of resonance come to figure oneself, as a dimension, as an inside and an outside, evoking the self to itself.[9] Accordingly, resonance is that force which brings space into the body and the body into space: "[sound] spreads through space, or rather it opens a space that is its own, the very spreading out of its resonance, its expansion and its reverberation." Within this flourishing, "space is immediately omnidimensional and transversate through all spaces."[10]

By way of such resonant conditions, one is figured by way of a greater movement, a type of expanding presence, and through which rooms and factories unfold as energetic and animate. Through Lucier's and Wollny's performative actions, space is compounded with resonances that come to evoke an omnidimensional construct, giving way to a sense for hearing beyond the apparent, that is, capturing the transformative acoustic potential that imparts a sense for the emergent.

Yet, resonance is also deeply troubling; within such sympathetic currents, such flourishing and transformative potential, resonance may also shatter—it may break form altogether, rupturing the lines that give shape to the body, to this room. Resonance, as Nancy also suggests, begins with a cry—the child's cry that breaks the silence, sets in motion a beginning: the beginning of oneself. If resonance is cast as a primary emergence, it highlights one as never fully individuated; rather, resonance shows the self as always in touch with a greater arena—an interdependent, social, and material arena: there is always something more, a "presence in presence" as Nancy states.[11] Resonance may put one in touch with oneself—it may evoke the self to itself, but it does so by figuring one as a spatiality, defined by a movement across inside and outside, and by extension, oneself *as* another; not so much collapsing distance, but rather, figuring one within a dynamic construct, or *omnidimension* of resounding contact.

Resonance, in moving across inside and outside, bringing space back to itself, evoking dimensionality—a continuous proliferation: a flourishing—as such, it is always already on the way to fragmentation, is held up as a state of transformation. This sounding, and resounding, this cry, figures "presence in presence," and what Nancy also terms "a complex of returns"—as what comes back to itself, yet always in excess of itself, always already *more than*. As a movement, a transformative state, resonance is radically unsteady, an oscillating dynamic of relation that, in trembling the limits of one's body, drawing out a connectedness, necessarily carries a potential of rupture, loss, disintegration, even erotic joy. Or rather, it integrates rupture as part of its material logic. In this sense, resonance is suggestive for an ordering principle that stays close to disorder; that weds attunement *and* interruption, sociality *and* escape, locating consonance and dissonance, harmony and discord upon its continuum. In this sense, resonance is always already disruptive of fixity: it pushes one away from oneself, completing oneself by way of relation, yet a relation to something else: something always stranger than imagined.

Returning to *I Am Sitting in a Room*, Lucier's work reveals more broadly how the movements of sound may be drawn out to unsettle or displace the delineations of form, animating material and spatial contours and, by extension, the facticity of things, to generate an emergent sense for the intelligible and the sensible. In short, poetic ecologies of resonance may contribute to processes of orientation, of knowing and understanding, that are equally about disorientation and dissolution: a generative loss of bearings in order to support a realigning or retuning of past and future, fact and fiction, oneself in relation to others. And importantly, to give room to the unrecognizable—*this that I may never recognize.*

Following these perspectives, I may emphasize the ways in which sound art captures and performs the dichotomy between phenomenological fact and sonic transformation less as a strict line and more as a resonant continuum; rather than withstand the incongruencies and discrepancies between fact and fiction, matter and dreams, works of sound art operate to *conduct* them. Consequently, sound art often maps out and directs us toward questions of cohabitation and coexistence I'm considering here. The crafting of poetic ecologies of resonance is founded on taking up residence within the unsteady assemblage of the acoustic and the acousmatic, the materiality of things and their resonant and volumetric energies—to assemble, as Wollny's work reveals, the told with the untold, the known with the unknown.

The work of Maryanne Amacher gives further expression to this quality, where sound functions to capture the specificity of places by explicitly drawing these out, beyond the immediate, locating facticity within a greater amplified construct. In particular, her *City-Links* series (1967–present) performs a dramatic articulation of an acoustically elaborated space, inviting one to inhabit its disorienting and reorienting qualities. Based on the live amplification of sounds culled from specific spaces and transmitted to other locations, and often incorporated into live performance or radio broadcasts, the *City-Links* series works to intensify a relation to material reality while diffusing such materiality toward a plane of "musicality." As the artist describes, "*City-Links* uses electronic means to connect spaces distant from each other. Together in time. Like music. Here there is no boundary to the outside. No wall to the outside. To receiving. But the wall might be with us. Inside."[12] Amacher, in linking as well as diffusing the fixity of places, accentuates sound as a medium for another type of site-specificity, another type of listening, one that upsets fixed dichotomies between physical matter and electroacoustic process, inside and outside, sound and music. In doing so, she captures and draws out a transformative, resonant potentiality: "My hope is that the split which now exists between these two worlds—that of musical language and of environmental sound—one day will be closed."[13]

Attunement and Interruption: Extra-Relationality

It is my view that works of sound art are often driven by experiences of *relational intensity*, constructing complex spaces and durations constituted by an encounter with distant locations, unknown energies, foreign bodies, elemental entities: the performing, recording, composing, and transmitting of animate and inanimate elements, energies and matters performatively aligned upon a mutational chain, toward a *being-open*, where the line between inside and outside shimmers, or closes altogether. Consequently, sound art may be understood to extend one's listening precisely where it's most difficult—upon those borders by which bodies and subjects, intelligibility and forms of recognition are often defined. As an art of listening—which is to say: experimental—works of sound art often bring one into contact with forms and forces of strangeness, bridging not only musical language and environmental sound, as Amacher poses, but also oneself and another, this human figure and other species, the seen and the unseen. It does so by testing the limits of listening, expanding the frames of address beyond that of language and human intelligibility; rather, listening is strained and stretched by an affective labor figuring an engagement with all that resides beyond human hearing and habitation: to relate one specifically to the borders of the almost perceptible, the almost recognizable, the almost tellable.[14]

Consequently, relationality shifts by way of the frequencies that pass across spatial and temporal coordinates, materializing in infra- and ultra-resonances; from the vibrating depths of the molecular and cellular to the radiophonic horizon of distant locations, from the unconscious and the fevers of recall to the material constructs of the social order, from the psychoacoustic to the electroacoustic to the bioacoustic, acoustic relationality is prone to complex expressions of coexistence: presence in presence in presence, a *being-with-with*. From such complex attunements and strange interruptions, poetic ecologies of resonance are supportive of the mutational: a type of generative *monstering* into which one moves or is moved.

The sociologist Hartmut Rosa argues for a greater consideration of resonance, which he identifies as an extremely enriching social capacity and experience. Resonance, for Rosa, is what binds us within an affective, interpersonal exchange with others, and which importantly gives way to a tangible sense of shared fulfillment. Through the flows of resonance, one nurtures a greater feeling for the value of social connection. While resonance, as Rosa outlines, may grant a deep sense of connectedness to the world and others, figuring a "sociology" of being in the world by way of vibrancy and responsiveness, I'm also moved to understand resonance as what transforms relationality itself.[15] Rosa suggests as much, in describing that sense of "collective transformation" that arises within scenes of mutual affection, and the bonds that emerge between people enacting together.[16] Through such scenes and experiences of connectedness, which opens us to each other, one may also begin to witness or sense how relationality is often more than what we imagine. In short, I'm led to wonder: How far might resonance take us? To what degree might resonance, as that which connects, and which is driven by a logic of transformation, put into question relationality itself as remaining bound to "sociology," to understandings of sociality? As I'm keen to consider, resonance may form the basis for exceeding relationality: where resonance becomes *restless*, a movement that keeps moving. Resonance, in this sense, may act as a tool, an affective knowledge, a medium of escape, for contending with the forces that often situate sociality in particular ways and forms.

Works of sound art draw out such questions by crafting material actualizations of resonance, figuring listening as the basis for radical connectedness—a raw relationality that is deeply experimental. Within the acoustic performativities and sonic constructs sound art produces, relationality is *strained* and brought to a resonating limit. Being connected to the world might be to follow resonance to a point where relationality is no longer recognizable, where openness is evoked and vibrancy enlivens toward the not-yet, or by way of the untold, manifesting an omnidimensionality that, as such, exceeds capture, or even meaning; for if resonance and vibrancy are to have affect, it must be on the level of placing one within a dynamics that exceeds what one knows of the world—*that hits me with an incomprehensible force, binding and unbinding me at the same moment*. Works of sound art may be understood to traffic in such a logic of transformation, where the limits of hearing become passageways toward undoing what constitutes bodies and things, engendering poetic ecologies of resonance—the bristling intensities of a geoacoustic world that, in moving us, always places us *somewhere else*.

In this regard, sound art often moves between cultural specificity and cultural ambiguity, relating the particularities of certain bodies, spaces, and histories (and their gendered, racial, classed conditions) to arenas of interruption and strange composition: what I may pose as a *mutational figuring*, a *becoming-resonant*, a restlessness, which might be the risk one takes in listening. In fact, I might ask: Where does identity go under the pull of an acoustic relationality and the transformative continuum of resonance? What becomes of Alvin Lucier through the operations of his mutational speech act, this flow of resonance that shimmers his voice, making it unrecognizable? Under the performative flow of Wollny's *Resonance Assembly*, what becomes of the disused factory in Berlin and its historical status? What becomes of place, its narratives or narrative-ability, within this sounded arena of affective knowing, this resonance that is always exceeding itself: this complex of returns that opens toward the untold, lending it a material shape (place in place, presence in presence)?

I'm interested to question in what ways an acoustic relationality—this art of listening—may afford opportunities for probing experiences and expressions of cohabitation and coexistence, as a possibility for elaborating the operations by which recognition is performed or fought for. The interweave of site and sitelessness, place and displacement, is captured as a generative framework for putting into question the lines of recognition, by calling upon a larger act of mobilization, a listening from below that may also emerge from above, from around or from within: a listening that works at acoustics as the basis for engendering a voice, a language, a sound art that may give traction to the unrecognizable, the unintelligible, and the not-yet told, the not-yet settled. In short, what I may image as an "identity politics" mutating under the pull of its own strangeness, or what Fred Moten suggests by way of musicality and "poetics"; poetics, following Moten, as "the differential resistance to enclosure," is what resists settling (gaining particular traction and significance within the context of the black radical tradition).[17] As such, poetics might be said to keep us within the resonant—not as what resolves, that closes or harmonizes, but rather, as what aims at connection; poetics as a *technique* of transformation—what may act upon things, resonating them until they come unfixed, to become partners in the radical reworking of the terms by which life takes hold.

It must be said that resonance, as an affective ground, a vibrant materialization of connectedness, is also open to instrumentalization. For instance, within the systems of neoliberalism, in which biocapitalism and the "experience economies" capture flows of affect, capitalizing on the vibrancies of life and sharing, resonance is put to work.[18] Or, we might detect a pronounced articulation of resonance in the emotional-infrastructures underpinning the "free market" whose transactions heatedly figure themselves by way of the "pulse" of global flows. In this regard, resonance is never always already free of capitalistic capture and instrumentalization; rather, as an affective ground it is susceptible, figuring one within any number of financial, social, and libidinal economies. Within such a view, resonance requires techniques, an affective knowing by which to stay with the potentiality of its excess, for how it may carry one in such a way as to challenge systems that work at settling the body, capitalizing specifically on one's capacity to feel, to move, to imagine, economizing the vibrancies that flourish by way of togetherness or restlessness. Resonance, therefore, might be more of a battleground, a micropolitics—an incredibly rich and dynamic capacity and construct requiring a critical move: to stay with the poetic excess that resonance inaugurates and carries.

From strange forms of speech and voice, language and expression, to diverse forms of listening and assembling, gathering around and tuning in, poetics may give ground for continuously inventing and reinventing ways of resonating: for weaving support structures *in the break*, that keep open and that tend to the open. As Nancy suggests, one is not only moved by way of listening, and the resonance it inaugurates, one is *forced* open. Poetics, following Moten, emerges from out of this being-open, this staying in the open, threading (and threatening) Nancy's "social ontology" with forms of radical exit or acts of unlikely solidarity.

Poetic ecologies of resonance—this omnidimensional, extra-relationality, this *monstering musicality* that resists settling—is posed here in order to nurture a particular framework or mode of listening, one I want to call, following the poetic vein, "listening-dreaming." If poetic ecologies of resonance, as I suggest, emerge as the meshing of phenomenological fact and sonic transformation, environmental sounds and musical languages, architecture and its energetic vitalities—*this room is no longer only what I see*—a continuum of sound that mutates as it goes, such poetic ecologies locate one upon the path of a mutational form of identity, a dreaming: *what am I really hearing within Amacher's City-Links?* What type of city is this now, and who or what populates this city, this linked urbanity? What type of citizen appears within these dirty compositions of urban excess? Runaways, resonant figures, dreamers—a listener that drifts, that is always somewhere else?

It is my view that poetic ecologies of resonance draw out and intensify what listening is always already doing: *relating me to that which I cannot relate.* For listening is fundamentally a listening *to others.* As Nancy suggests, listening *evokes* another, as that which can be listened to, even oneself, or this machine, this animal or insect, the ocean. Listening is the very thing that evokes relation: to something, an other. As such, following the dynamic constructs of sound art, I'm led to ask: How might one inhabit this complex and radically unstable weave of the facticity of things with the sonic experience of the unseen, these transmitted urban sounds that emerge from out of nowhere? How do we take up residence within the constructs of poetic ecologies to navigate through that which disorients and elaborates with such intensity and power? That upsets the lines of this body, this knowing, extends them, and resonates them, attuning one to the flux and joy of an energetic volume that is always greater than I may know—that mobilizes an affective unease and desire?

Listening-dreaming is thought here as a mode of listening that may navigate the here and there of acoustic relationality, which I may underscore more accurately as "extra-relational"; a listening that both attends to the immediacy of space, and what sound punctuates in the present, as a bringing to presence, while also relating to that which listening always opens toward: a here *and* a there. For if listening is what figures oneself as a being-with, as Nancy suggests, it requires a particular capacity: a sense for being here *and* elsewhere, underscoring specificity as always already interrupted (or fulfilled) by ambiguity. In this sense, I'm posing listening-dreaming as an active mode, even transgressive, by which to be here and elsewhere, for dreaming is that form of attention or cognition relating to what is *not there*, or not there any longer, and which may also never appear, but nonetheless figures itself, is present, somehow, and persistently, within the present: the unconscious figuring, this drifting imaginary, the not-yet told coming out, presence in presence. Dreaming as what starts to give narrative precisely to the traumas, joys, hopes, fantasies, those strange ideas, and which drags one along in its complex expressions. Listening-dreaming may be considered as a mode of extra-relational listening, one that may follow the mutational chain of sound, allowing for the unknown, a *not knowing*, and that may assist in taking up residence within a poetic ecology. If listening opens oneself, if it forces open by way of resonance, it must open equally toward that which one never fully understands.

Toward the Unrecognizable: Watery Life and Worlds of Matter

In order to elaborate questions of acoustic relationality, and the mode of listening-dreaming I'm posing, I'll turn to a number of practitioners of sound art, as well as tendencies defining the field. This may allow for probing further the relational challenges and potentialities found by way of resonance and the poetic ecologies it inaugurates.

The recording practice of the artist Jana Winderen, for example, deepens a view onto the rather hidden regions of oceans and rivers, capturing the sounds of underwater life and the complex murmurings of all that lurks below. In this gesture of recording, we witness an essential activity found within sound art, that of casting one's listening toward the unfamiliar and what lives deep within or beyond human habitats. Mostly based on dropping a hydrophone device (designed to detect and record underwater sounds) into rivers and oceans, from the glaciers in the North Sea to rivers in Thailand, one encounters, through Winderen's recordings, the force of an altogether foreign environment; from shrimps to urchins, water currents to industrial vibrations, Winderen is fascinated by and concerned about the hidden depths and unseen worlds of water. As such, her works immerse us within a poetic ecology in which to listen is to confront that which is never fully identifiable—*what are these sounds one hears?* Is it possible to ever know for sure, all that these recordings contain? From scratched textures to mesmerizing crackling, tonal swooshes to deep murmurings, the underwater environments resound as complex soundscapes of the familiar and the unfamiliar, the known and the unknown.

Winderen's recordings, and subsequent compositions and installations, while considering the territorial qualities of the oceans and the seas, the conflicts passing between life underwater and the human industries and labors that condition and interfere with such life, do so by inviting one into an affective encounter, a crafting of poetic form by which to extend documentary capture, exciting its methods, its objectivity. While referring to a longer history of ethnographic field recording, and the recording of wild life, I'm interested how within the field of sound art such an approach is driven not necessarily by a desire to find out something, but to mobilize other entry points, and other experiences. These, rather, are shaped by a focus on listening, and give way to articulating a listening practice over a strictly ethnographic one. Listening might be appreciated as the motivating factor, prompted by a concern for all that is out there, and which listening is particularly poised to address.

> 21st April 2015, a trip to Camp Barneo and the Geographical North Pole: made recordings 15 meters under the ice with hydrophones. The sound of the moving ice could be discerned—a beautiful and intense sonic experience. To walk on this drifting skin—this sea ice cooling the pole—is an intensely moving experience; by communicating this through sound compositions based on the recordings taken from the ice itself, should evoke concern, respect and curiosity. By communicating with sound the reality of a place, the different issues become more intense for the audience than if other media were used, for example photography or other image reproduction.[19]

As Winderen notes, listening as a deeply generative approach underpins the experience of the artist as well as that of the audience—listening as what takes shape, emerges, and gives way not so much to "capturing," but rather, to evoking an intensity—of "respect and curiosity": a *dwelling alongside* the (life)forms and forces the recordings trace, amplify, or compose.[20]

I'm interested and fascinated by Winderen's approach and practice, where listening extends itself toward the unseen depths, specifically intensifying contact with what is *out there*. Her works attend to lifeworlds found in habitats radically different from one's own, allowing for the distant and the strange to become proximate and hearable. As the artist highlights, she mostly does not know the origin of what she records, and yet, understands these sounds as expressive of a world of life as well as conflicting forces.[21] Through such listening experiences, one encounters the unrecognizable, as that which is beyond full legibility and knowing—a literal depth that confounds knowing with its vague contours. Rather, as a listener, a *listener-dreamer*, one is situated within an "intensity" of relation, where the force of *other life* touches one's own. Does not the brimming flow of these sounds remind one of the planetary ecologies of which one is a part? Turning the volume up, letting these watery sounds fill the room, one is overtaken by the intensity of not knowing and yet knowing something, the power of these sounds into whose composition one is equally stitched: for these recordings and works draw one's own breath and vitality into the movement of others, these creatures, these currents to which one is also bound, giving way to a *creaturely resonance*, that may allow one to take up poetic residence alongside and with the aquatic currents and marine life defining much of the planet.[22]

I take such sonic work as a form of aesthetics that researches the particularities of an acoustic relationality, where sound works at forms of sensing and cohabitation that exceed one's own body, the human world, and the world of intelligibility; a research onto what's at stake in listening. Listening as what may traffic across a range of acoustic sites and scenes, affording "planetary" connections, figuring resonances across a diversity of situations, species, and modes of life. In the case of Winderen, this may bring forward a scene in which one may follow listening as the basis for staying with a world so distant and yet so close to one's own.

Extending from Jana Winderen's recordings, I want to consider the practice and methods of the artist John Grzinich, who also works through processes of fieldwork as well as sited improvisation. Often focused on individual or collective sonic improvisations within peripheral or ruinous environments, Grzinich has been making recordings for a number of years based on interacting with a disused water tower near his home in Estonia, resulting in what the artist calls "portrait of a sounding object."[23] Bowing and tapping the object, with microphones placed within or directly on it, Grzinich draws out an array of textures and tonalities. In doing so, the recorded portraits exemplify a method of sonic site-specificity in which found materials and architectural structures function as instrumental opportunities—as carriers of possible sound. I'm interested in how the act of sounding the water tower draws out a view onto a form of industrial detritus, and a particular industrial, historical narrative, by also shifting focus, moving closer to touch its rusty surface, tapping its materiality. In doing so, Grzinich also estranges one's sense of knowing, recasting the particularities of this object by way of a resonating, auditory act. From the distant to the proximate, from a historical context to the materiality of the thing, such perspectives are held together in the resonating tonalities, the sonic composition. Such actions articulate a sense for collaborating with the world of matter: the material things that figure so prominently around us. As Grzinich outlines:

> Having access to more information is helpful for environmental research as the systems we need to study are extremely complex. On the other hand, the more we know, the more we realize how much damage is being done to the delicate webs of life at the expense of ever expanding human activity. The risk is that, future generations will feel ever more compelled to preserve what little is left and devote too much attention to reconstructing the past. I've seen this with folklore and music that relies on a limited set of photos and sound recordings, around which whole movements and even national identities are constructed. I feel the focus of my work has been more about developing an awareness of the environment through what we sense, so that we can live more in the present than in the past. Recording capability is more of a tool for building relations, than a method for preservation.[24]

Grzinich's approach underscores the importance of listening as a practice, and how the activity of recording enables ways of engaging with the matter of environments—for "building relations." As with the musical instrument, the materiality of things is not so much inert, rather it comes to figure one's body, one's touch by way of its particular shape, texture, tonality, giving way to a pronounced sense of co-creation: one is played by the object as much as one plays it. I follow Grzinich's work, his sounding portraits, as portraits that are more collaborative than individual; portraits articulated by an attentiveness to the present, a resonant engagement that acts to intensify relations to one's material surroundings as a scene of discovery and encounter, a way of knowing.

In parallel to Winderen's field recordings, as well as Lucier's *I Am Sitting in a Room*, Grzinich's approach underscores how practices of sound art often search for ways of reorienting one's sense for the conditions defining environments, thereby instigating new alignments and synchronizations—these instances of fieldwork and improvisation, and the extended resonances that reorient a sense for the specificities of environments and their histories. Accordingly,

I may begin to outline acoustic relationality as one that exceeds the relational as being bound to intelligibility, to human relations and understanding, and processed or supported through language, discourse, and the operations of a cultural specificity. Rather, acoustic relationality is an affective labor—a type of research or activism aiming to resituate one's sense of knowing, and resulting in more embodied, tactile, and collaborative encounters with other things and life-forms, matters and energies.

Practices of sound art work at amplifying specificity, the facticity of things and places, to a point of intensity, even *breakage*, opening out to possibilities of engagement with what also often eludes engagement; a transformative or generative logic that requests another form of attention, here and there. As such, we may speak of *extra-relationality* as being founded on the interweave of material fact and sonic transformation; a poetic ecology of resonance that comes to articulate an excess of relationality—a not knowing for sure, and that may find narrative through other means, a listening-dreaming that may articulate a sense for what exceeds understanding. Extra-relationality attempts to augment the experiences of being-with underpinning subjectivity with the vitality of contact that is always exceeding the limits of a social context: other voices, other matters, and other energies that move in and through the body, and that may also leave my body behind. For resonance surely bypasses and exceeds the realm of human sociality, of cultural specificity, to integrate into its omnidimensional proliferation all sorts of matters and energies, voices and entities, making of one's body a strange composition: a *commotion*. Extra-relationality as a defining quality of the poetic ecologies I'm outlining comes to pose listening as an act that engages more than the hearable, that hears beyond itself.

Poetic Cosmos: Dream-Action

The works of Jana Winderen and John Grzinich return us to the critical relation between material facticity and sonic transformation, and how this comes to define a conceptual territory for sound art—the desire to collapse the walls between music and environments as Amacher states. Winderen and Grzinich are exemplary of certain approaches, where acts of sounding and listening, touching and recording engage with what is to be found in distant and derelict environments, bringing this into a space of relational intensity—a listening practice.

One may detect within works of sound art a drive toward dissolution and discovery, a desire to be interrupted by something outside oneself, which may yet assist in reworking one's sense of orientation and situatedness. In this regard, what becomes of situatedness within poetic ecologies, where a logic of transformation, of emergence, constitutes one's subject position and the understanding of situations? What forms of responsibility or responsiveness are required or shaped by the resonant complexity of the extra-relational?

Here, it is helpful to recall the ways that Alvin Lucier's *I Am Sitting in a Room* is a work that not only explores and composes an acoustic (and technical) process, resulting in a musical work, but also is equally driven by the composer's desire to overcome his own stutter, or what he terms the "irregularities" of his speech. As Lucier states: "I regard this activity not so much as a demonstration of a physical fact, but more as a way to smooth out any irregularities my speech might have."[25] Through the processual unfolding of the composition, his speaking and recording, and re-recording, Lucier works at reshaping the defining quality of his own voice, this resonance that subsumes his voice. In short, he becomes something other to himself by shifting the break in his speech, this disfluence, into that of a shimmering flow of sound. In this sense, *I Am Sitting in a Room* accentuates a potentiality particular to sound and experiences of listening, one that may tend toward states of dissolution, breakdown, as well as reconfiguration, even repair, a mending.

Lucier's composition asks of us to bear witness to a particular transformative act, one equally shaped by a temporal structure: the work is only constituted by an insistent relation to a past, a past that never fully leaves as it is replayed, again and again, until the resonant frequencies of the room reinforce themselves, as well as carry forward a future in the making, toward a mutational body—prompting the question: How far may resonance take us? This future is already announced in the work, as the promise of overcoming irregular speech, and by extension, the thingness of the body and the room itself, as material, phenomenological facts—as a situation. Rather, the materiality of the room, and the corporeality of the composer, transmute into "music," into resonance—"the complex of returns" Nancy highlights, and whose refrain imparts another manifestation of space and speech, and by extension, inhabitation. Where is Lucier by the end of the recording? Is he still in the room, and is the room still the same room? What have we, as listeners, in fact witnessed or experienced? If Lucier were to continue, extending past the thirty-two repetitions, where might we end up?

Gaston Bachelard, in his book *The Poetics of Reverie*, develops an in-depth account of daydreaming, which may be mobilized to reflect more fully upon the specifics of poetic ecologies I'm outlining. For Bachelard, the path of reverie is a "downhill path" where consciousness wanders and becomes clouded; it is an instant of in/attention, giving way to a sensing precisely of what hovers around and within the materiality of everyday life: an *oneiric dimension* in whose emergence the imaginary and the real are joined. As Bachelard reminds, we keep our eyes open when we daydream. Subsequently, there exists a double-perception, a ghosted sensing: on the one hand, we perceive what is in front of us, a table for instance, or people walking by, and yet, one becomes additionally fixated upon something else, something that arises from the scene but also hovers within one's thoughts—a distracted thought, a *dreaming along with* that necessarily shudders reality to grant "the real object its imaginary double."[26]

Importantly, for Bachelard, reverie acts as the basis for the emergence of the poetic imagination, a constructive, creative drive of *becoming unreal*: a dissolution that enables a flow of thought that is always aiming for other worlds. As Bachelard poses, in its newness, its shimmering appearance, the poetic image "opens a future to language."[27] Reverie educates on how to dream during the day; to pull from the banal and the ordinary, from language itself, an extraordinary event or expression; to find upon this downhill path a means for relating to the future. The poetic imagination attempts to take root, to figure itself in the world, as a type of *unreal reality*. In this regard, the daydream, in tracing over reality, reveals new directions and dimensions for inhabitation: to being in the world yet as found along the "wrong path" of reverie.[28]

Reverie is thus a type of pause in whose gap the real and the imaginary come into contact, existing side by side, or one over the other, as a dynamic, unstable composite. Reverie comes to locate the real and the imaginary into a form of composition, giving way to a "double consciousness" destabilizing the borders of the animate and inanimate.[29] This composition, this intertwining of the real and the imaginary, Bachelard ultimately refers to as the making of a "poetic cosmos," a cosmos arising by way of daydreaming. As Bachelard ultimately suggests: "Poetic reveries situate us in a world and not in society."[30]

A poetics of reverie opens a path from which to begin to elaborate understandings of the extra-relational that are deeply enriched by the seen as well as the unseen, and that situates the legible and the articulated as always already tethered to the imaginary. As a mode of dreaming, reverie wields an inner work that finds its strength, its enduring passion through resonance, a sympathetic flourishing that brings the inner and outer into shared animation—as the shimmering of form and meaning. Thus, reverie is a particular form of attention and experience often equated with the poetic, as that creative tuning of self and surrounding. The poetic might be said to materialize as a *writing while listening*, one that is always *opening out*, following the emergent flow of connections; a *writing aloud* as Roland Barthes offers, giving way to a textual bliss.[31]

Reverie is that special form of dreaming that nurtures poetic imagination and that draws out a resonance on the page, that composes from such ephemeral and restless moments, and by which connections proliferate, *to be always already here and there*. This may find an echo in what Pauline Oliveros terms "quantum listening," as that state of listening able to hear across multiple levels and scales simultaneously, from the micro to the macro, the terrestrial to the spectral.[32] Listening is called upon as a capacity for being in touch precisely with that which is beyond human scale, beyond the fully quantifiable. Such a notion parallels Bachelard's understanding of reverie, and is suggestive for further outlining listening-dreaming as a particular mode or form by which to inhabit poetic ecologies—these ecologies that are constituted by connections across the micro and the macro, facts and dreams.

This finds a compelling expression in the sound works of Christina Kubisch, in particular her "electrical walks."[33] These are based on utilizing custom-built headphones that act to receive electromagnetic energy in the environment. Walking through cities, along streets and into shops, one receives and hears electromagnetic energy coming from all the electronic, cellular, and network signaling passing through our environments. From mobile phone signals to security alarm systems, from video game arcades to subway trains, from the micro-fluctuations of our smallest devices to the macro-structures that facilitate and manage infrastructural systems, Kubisch attunes us to the disturbing, often strange and beautiful world of energy. Walking through the city, one enters fully into a state of listening-dreaming: one sees the material world around while hearing a parallel world, a world of signals that is radically overwhelming, confusing, disorienting—one can never grasp the full complexity or dimensionality of what these signals reveal. Rather, one is called into an impossible listening, a poetic dreaming by which to navigate through this airborne ocean of electromagnetic force equally defining, in an extremely invisible way, much of the globe.

Although Bachelard focuses on a poetic cosmos of daydreaming as something that takes one out of society, I'm interested in the idea of *poetic subjectivity*—this listener-dreamer, this electrical walker—that specifically enables forms of social imagination: that wields great influence over the ways in which we may perform as social subjects. In an essay on the politics of resistance, Elena Loizidou lays out a critical framework for approaching "political subjectivity," one that importantly integrates individual dreams and their telling. As she proposes, "If we exclude dreams from our accounts of political subjectivity, we fail to account for the singularities of political subjects and not just the political potential residing in these singularities."[34] For dreams are fundamental to how we understand ourselves as subjects, as persons, and how subjectivity is entangled within the complex weave of conscious and unconscious worlds and experiences. In steering us toward dreams, Loizidou offers insight into the ways in which political life is shaped by desires, wishes, fantasies, as well as complex histories and memories. Dreaming, therefore, is posed as a form of action, a "dream-action" as she suggests, through which realities are often reimagined, or from which one may work through the traumas of injustice.[35]

Subsequently, one may appreciate how the generative formation of a poetic cosmos may give way to a sense of empowerment, linking the fears and desires, memories and hopes of individuals with the imaginative drive of how these might be worked out—how the facticity of the world may be transformed. And importantly, how we may come to hear, in the narrating we give of ourselves and each other, the meaningful figuring of a transformative act. In this sense, listening-dreaming might best be described as a form of *double-listening*. A double-listening situated along the here and there of acoustic becoming; that captures the oscillating dynamic of poetics, as what shimmers the lines of things with radical imagination. A listening that may move by way of memory and history, and toward a future to come. In this regard, I want to turn Bachelard's poetic cosmos of daydreaming toward ecologies of resonance, as the intertwining of dreams and matter, the lines of the body with the potentiality of its reconstitution, its reconfiguration within the sphere of the social. The poetics of listening-dreaming might be mobilized to figure the social as always already interrupted by the unrecognizable, a sociality in excess of itself, that mutates as it goes, shimmered by the complex of returns, or by the recurring dreams that refuse to leave us.

Such mutational perspectives can be understood as the poetic move from sense to nonsense, where poetics is aligned with the invention of new languages: a speaking, a writing, an expression that stretches meaning and the rules of grammar, syntax, even that of the phoneme. All such matters of language are reworked in the poetic mutational dream-act, this resonating imaginary. Poetics is a force that moves language, and as such, it overturns the lines separating sense and nonsense, and by extension, the intelligible and the unintelligible, with all sorts of strange images, broken phrases, mistranslations, excessive wording, shimmering lines, a festivity of ideas. Sense and nonsense are thus agitated and interwoven, figuring what may be considered as *extra-sense*: an excess of sense. In short, poetics acts upon the inherent excess of meaning embedded within language itself, capturing and refining the "madness" already lurking therein.[36]

Bringing Bachelard's cosmos back onto the sphere of the social, is to also figure a force of the antisocial: for in the invention of new language, however guided by the vernacular music of the social, one travels by way of an anti-sociality, a restlessness, escape: this language that is also cosmic, that turns alien, and that *flees*. Poetics might be said to intersect the social and the cosmic, sense and nonsense, drawing these into a strange continuum: that moves, moves wildly, and in doing so, captures *everything* in its movement. From "dampened thunderbolts and dirty lightning flashes," to that of "a dove or a poppy or a kiss," and further, where "vowels are drowned" giving way to "violins filled with water, fresh explosions, submerged motors and dusty shadow, factories, kisses, throbbing bottles, throats." In following Pablo Neruda's mesmerizing flow of poetic images, language is prodded into movement by way of resonance—the resonance, which as Neruda captures, makes the day "stand out." As Neruda writes: "To resonance the soul rushes / and celebrates and hastens its swift wedding."[37]

Following Neruda is to follow resonance as it draws the day out, brings it forward to itself, celebrating the wedding of matter and energy, the marriage of the facticity of things with the rushing of the soul, from "silence" to "instant roses." Poetics is thus poised as what honors the rush of life, putting such a rush into language and meaning. By way of *extra-sense* articulations, a language brimming with sense, poetics also invites one to inhabit another form of sociality: to stay close to the limits of the recognizable, the sensible, moving those limits by way of new languages, and new connections.

If sound, and the transformative acts I'm mapping, is that particular medium or matter able to nurture passage between fact and fiction, work and dreams, things and feeling, it provides a deeply expressive means for animating hidden feelings or repressed memories: for probing and engaging that which remains unspoken, punctuating frustrations as well as desires: an act of resounding in order to return to oneself, and to transform along the way—to support the *rushing of the soul*. As found in a range of acts and actions, sound potently enables the expression of angst and frustration, love and passion; whether in the musicalities of house parties or community struggles, the cacophonies of street demonstrations, or the rapturous vocalizations of the forlorn and the injured, the thrilled and the passionate, sound punctuates instances of loss and fragmentation, erotic joy and social struggle, regret and hope in the future. Such sounding out, such sonorous outpourings and amplifications are figured as enabling processes of resistance and determination that are, at the same time, renewing and repairing. Sound, in this sense, operates as a carrier, a continuum, for individual and collective transformation; for working at *opening up* that which is locked down, held back, or that which simply needs expressing. Such tendencies and potentialities, though, are enabled precisely by sound's capacity to unfurl a particular imaginary, a critical dreaming, a dream-act as Loizidou outlines, which wields a transformative logic—a sense, or *extra-sense*, for reorienting what binds one to the present. That is, a means for negotiating or celebrating the particularities that hold us.

Poetics on the Move: Sonic Fictions

I'm working through a range of perspectives onto questions of relationality, dwelling upon the poetic figure of resonance in order to suggest ways of approaching what being-in-relation might come to mean. Acoustics, and the practices of sound art, becomes suggestive for bringing forward a certain dynamic onto this questioning, for underscoring recognition as a politics and a poetics, which requires continual probing, an activism, an imaginary to put into question what it means to be recognized and what forms recognition might take. Here, what comes to define being-in-relation, as the articulation of a social community of equals, must also contend with the forces that define the nature of that equality, the social or discursive frames through which mutual recognition is understood and articulated.

In seeking to mobilize Bachelard's poetic cosmos, suggested by way of Loizidou's dream-action, I may find a point of productive dialogue with the writings of Paul Gilroy, and the ways in which he charts out the movements of culture across and by way of the black Atlantic.[38] As Gilroy argues, the black Atlantic captures a history defined by coloniality, the brutal realities of slavery, and a subsequent genealogy of passage and racial struggle. Consequently, it additionally speaks toward the ways in which such history contributes to the figuring of transnational or "fractal" identities and cultures. Defined by way of passage, the black Atlantic is founded on racial violence that instigates the arrival of complex identities and that force into the history of modernity a set of radical refusals and cultural expressions. As part of this complexity, as Gilroy outlines (by way of W. E. B. Du Bois), a "double consciousness" comes to shape the life and expressions of black Atlantic culture—a "looking both ways" (across the Atlantic, across cultural homes, across historical ruptures) from out of whose performance constructions of identity gain a fractal complexity and shape.[39]

Charting the particularities of the black Atlantic, Gilroy works at upsetting articulations of nationalism and national identities defined by way of "racial purity" as well as ethnic authenticity; rather, by way of a black Atlantic perspective Gilroy figures a modernity always already transcultural, unsettled, shaped by way of the literal and figurative transit of "ships" (which the author critically deploys) perennially passing between continents and homes. Ships are captured, or *pirated*, within this fractal perspective of "looking both ways," as "cultural and political units" whose mobility between cultural nodes engenders particular knowledges and expressions, languages and imaginaries born from the conflicts and diverse contacts defining the black Atlantic.[40] In tracing the specific and speculative routes charted, from across the Atlantic and beyond, Gilroy attends to the musics, performances, and sounded cultures and guises that emerge, attuning to the cross-fertilizations and transfigurations arising between and across nation-states and ethnic homes. In doing so, he forges a critical method, a creative listening, which captures the particular as well as wayward paths of modernity and its subjects. At stake for Gilroy though is not only the capturing and description of hybrid musical forms and transcultural expressions, but more, to deploy a musical and performative criticality against the reign of "textuality" underpinning modern liberal democracy and its bourgeois public sphere, especially when such textuality may enclose black identity within colonial or even postcolonial terms. By way of black music traditions and expressions, Gilroy argues for the necessity to attend to the felt experiences and embodied, sonic knowledges of black musical expression and performance as a driving force for redefining constructs of identity, citizenship, and community, and how we come to understand the intellectual or moral force of public discourses.[41]

The issue of double consciousness provides a productive ground for getting at as well as complicating the questions and topics I'm engaging. Gilroy's work on black Atlantic culture usefully impacts onto the transformative understanding of listening and extra-relationality I'm addressing by way of sound art, and may allow for mobilizing a number of questions, for instance: In what ways might an acoustic politics, and the notion of poetic ecologies of resonance, allow for articulations of social solidarities that challenge nationalistic movement and its quest for authenticity? In what sense might the poetics and politics of black Atlantic culture and thinking contribute to capturing the dynamics at work in the extra-relational found in listening beyond the familiar, and for inhabiting the particular resonant journeys such listening incites? To consider the extra-relational as a form of dispossession, and as something that must contend with genealogies of violence? What critical and creative tools might be found from within diaspora communities that can assist in navigating the gendered or ethnic discourses often defining home and identity, and that may give a certain critical impress onto the dream-actions of political subjectivity?

Paul Gilroy's reflections on the double consciousness of black Atlantic culture impart a critical base for understanding the ways in which diaspora communities configure themselves in relation to a past that is no longer fully accessible and a present that is never quite home. This doubleness is thus understood as a complex cultural and psychological position that may give way to questioning and negotiating the frames placed around particular subjects which, as Gilroy suggests, often "textualize" one as being bound to particular national, political, or ethnic narratives. By way of the affective tensions, symbolic narratives, and social attunements underpinning musical performance, for example, certain capacities emerge that may wield their power in an effort to stretch a public sphere and what constitutes its discourses.

Following such perspectives, listening is mobilized as what may attune to certain situated meanings and stories while equally aiding in figuring an emancipatory potentiality. A listening that doubles itself, that turns this way and then that way, and that joins itself to the movements of a certain collective imagination. A double-listening is imaged then as the work one may do in order to reconfigure the status of the legible and the illegible, between the frames that situate the voice and its specificity, and those that may give way in support of other soundings and other passages. A double-listening that navigates the complex of returns—of turning this way and then that way, where inside and outside, past and future, tether one to certain continua. What might emerge, then, is not necessarily a flight from the binds of daily life; rather, a new sense for what it might mean to be situated. A double-listening asks of us to hear more, to listen again, to listen both ways, to keep on listening in order to traffic in an altogether different movement aimed at detuning origin and exile in favor of the mix, a poetics.

Questions of listening, and the mix that is culture on the move, find critical traction in what Kodwo Eshun terms "sonic fiction," which he elaborates by way of all types of funk, that electric audio-verse of stylized sonic-cosmic mixing. From the funky fictions of Parliament to the disoriented dub-induced studio work of Lee Perry, sonic fictions deploy a range of worlding narratives that twist our listening toward that of alien or mutational life. Eshun's sonic fictions are shaped by a refusal to subscribe to "today's journalism" which, as he states:

> is nothing more than a giant inertia engine to put the brakes on breaks, a moronizer placing all thought on permanent pause, a futureshock absorber, forever shielding its readers from the future's cuts, tracks, scratches. Behind the assumed virtue of keeping rhythm mute, there is a none-too-veiled hostility towards analyzing rhythm at all. Too many ideas spoil the party. Too much speculation kills "dance music," by "intellectualizing" it to death.[42]

In response, Eshun guides us into the dizzying flux of all rhythms, unpacking the "Mother Rhythm" which may fundamentally push us into step with what listening may stake out: to relate to the trodden paths of particular genealogies by driving at the ever-emergent sounding of an alien future.

The articulations of diasporic sound and listening, and the itinerant oscillations embedded in sonic fictions, are creatively charted by the artist and musician Satch Hoyt. His Afro-Sonic Mapping project captures the concept of sonic fiction Eshun poses, drawing out questions as to the mixing of musical and cultural signifiers.[43] Hoyt's project looks toward histories of the black Atlantic by focusing on the migration of "sonic signifiers," in particular African rhythms that, along with the violent enslavement of Africans, were transported and transplanted throughout the Americas and beyond. Taking as his starting point a collection of anthropological field recordings of African music made in the late nineteenth century, and mostly held in the Phonogram Archives in Berlin, Hoyt travels to some of the original sites and cities, such as Luanda, as well as to selected destinations in Europe and South America. Bringing the recordings with him, the project develops as a series of interviews and collaborations with musicians, cultural researchers, poets, and historians, in order to listen and compose music together, reflecting upon the recordings as sites of colonial capture as well as resistance and creative endurance. Hoyt is interested in considering historical injustices, sonic migrations, and personal journeys by understanding the passage of "the Afro-Sonic Signifier" as an ongoing narrative of black culture. Such a concern is manifested by finding resonances between the historical recordings and contemporary music cultures. In this sense, Hoyt underscores the migrations of sound as a complex story, one that gives support to (post-)colonial struggles through the musicalities and sonic temporalities that speak toward the homeland as well as distant worlds, and that never cease to travel, where rhythms, in particular, continuously re-emerge in music far from home.

What Hoyt captures and prolongs are the ways in which listening is situated within a weave of references, histories, desires, and cultural signifiers. Afro-Sonic Mapping is not only a map of the "eternal signifier" of African rhythms, but one that equally figures a poetic ecology, a sonic fiction, where the resonances across time and space oblige us to hear differently: to work through the mix that is black modernity as well as that of the unhomed self inhabiting this globalized world. Importantly, rhythm surfaces as the key signifier, the fundamental carrier of cultural expression that beats out particular meanings. While resonance may be found in making connections across time and space, rhythm puts resonance into particular grooves, impressing or importing meaning onto its vibrant continuum by way of the force of certain cultural moves.

Following questions of sonic fiction, and the double consciousness mapped by Gilroy, works of sound art more generally can be appreciated as fueling experiments in working through the intersections of place and displacement, addressing states of being unhomed—of being defined by a complex of returns. From seeking out distant locations and nonhuman entities, linking cities together in a web of amplification, or recomposing one's voice as a diffuse articulation of resonance, sound art is deeply concerned with states of dislocation and displacement. These, in fact, become strategies, methods, forms of listening. Yet, in a way that reorders their conditions, positing such artistic expressions as allowing for another sense of belonging and participation: to find or rework understandings of home within states of disorientation. As Gilroy suggests, music might be heard to dramatically enact forms of support, figuring a sonic fiction which grants shelter against experiences of trauma and rupture by explicitly spiriting a sense for transformation, for another type of future.

Listening-dreaming is positioned here as a power, one of attention and poetic sense, a *double-listening* that guides one in and around the apparent and the legible, unpacking the archive as a reverberant body, to reorient the here and now by way of strange matters and voices. As such, poetic ecologies of resonance are set in motion to redefine constructs of relationality as often being shaped by recognition and intelligibility, by that which appears, named or nameable, *textual*. In doing so, I'm curious in what ways such ecologies, and the transformative or mutational logic they incite, may support a political tussle with the normative lines defining recognition. Here, I follow Athena Athanasiou and Judith Butler in their discussions on dispossession and the politics of recognition. As they suggest, recognition, while extremely important, may inadvertently operate as a form of "capture" by rendering the unrecognizable intelligible according to the normative powers of particular institutions, communities, languages, or histories. As they argue: "The point is not to clamor for recognition at all costs, to conform to the schemes of intelligibility that register as assaultive, but to examine the costs of recognition with the struggle for survival."[44] As such, it becomes imperative to not only lay claims for recognition, but to do so in a way that puts into question the structures that define the lines by which recognition is granted. In short, one must ask: Recognition according to what or to whom?

By drawing into question the issue of intelligibility, and a politics of recognition, Athanasiou and Butler provide a framework for exploring in what way recognition can be creatively and critically performed, worked over, or destabilized—in short, to approach recognition and the drive toward intelligibility as an endeavor that must additionally impact on its defining logic. "Recognition itself has to be a transformative category, or it has to work to make the potential for transformation into the aim of politics."[45] In this regard, I'm interested in the ways in which an acoustic relationality—the extra-sense of a listener-dreamer that looks and listens both ways—leans upon the interferences and rhythms offered by figures and forces of strangeness, by the inanimate and the animate alike, to give way to a movement, a being alongside others. And as such, puts recognition on the move.

As previously discussed, the work of sociologist Hartmut Rosa aims at extending understandings of recognition by posing resonance as a type of supplemental notion. In dialogue with Axel Honneth's work on recognition, which underscores the importance of recognition as imparting a sense of individual worth and self-esteem, Rosa argues for resonance as what may "extend" recognition. Resonance, according to Rosa, is a "dynamic event" and the "expression of a vibrant responsive relationship" that exists between people, and between oneself and the world.[46] While recognition can be fought for and granted, through a range of struggles and forms of institutional implementation, in the granting of legal rights for example, resonance is a living process of "mutuality" that can lead to a deeper sense of self-esteem and self-worth. Recognition and resonance come to form two intertwining positions within a framework of relationality and justice, yet it is by way of resonance that Rosa points toward a more dynamic sense for extending from where recognition may conclude; where resonance necessitates that one remain open to the world and others, an openness that may enliven one's sense of recognition by always already extending that which connects us, that fulfills us.[47] In this sense, resonance might contribute to Athanasiou's and Butler's call for recognition as a transformative category, as what impacts politically as part of its realization. By considering resonance as an extension of recognition, as Rosa suggests, as being generative of a more pronounced form of attention to others that figures a transformative space, as such it may be understood to move recognition and social equality closer to a condition of emergence: as something responsive to the complexities of identity and how we engage with each other. That is, resonance may support precisely where recognition is withdrawn or foreclosed, as part of the passages some identities are forced or asked to make.

Acoustic relationality, as I'm outlining, is underscored as the basis for poetic ecologies of resonance, which exposes one to the "vibrancy" Rosa describes as well as the complex performances Gilroy underscores: these "fractal" identities and cultural expressions that navigate, by way of musical matters, states of passage. As such, acoustic relationality is put forward as a critical base for questioning the pull toward recognition as what may claim differences by integrating them within a frame of legible capture, within particular narratives. Rather, by integrating a sense for movement across site and sitelessness, the facticity of things and their transformative potential, by arguing for spaces and archives as zones of passage, as dynamic events, by hearing more than one can ever fully recognize or know, intersections of the distant and the proximate, the human and nonhuman, identity and its mutational future, are provoked.

Stranger Publics: Songs and Screams

Through a consideration of expressions of sound art, practices of sounding and listening may be highlighted as a means for engendering a sense of the extra-relational, one that extends or interrupts the dialogical and the intelligible, the recognizable and the culturally bound. I'm interested in the ways in which such practices produce relations shaped by a poetics, an imaginary, a dream-action that may narrate memory and history, granting a futurity by embedding them within the affective labor of listening: a resonance that may figure a critical tussle with the frames of recognition and legibility. As sound art highlights, listening may be underscored as enabling for an orientation toward the unseen or unidentifiable: even one's voice that may become unrecognizable. Sonic fictions, and the fractal constructs of black Atlantic culture, help ground the question of poetics and resonance by locking them into particular steps. As such, they bring into poetic ecologies the question and temporal force of rhythm; by way of rhythm, time is figured as material, contributing to the transformative logic of acoustic propagations, a temporal dynamic. While resonance moves space, bringing space to itself, rhythm materializes time, granting energy and shape to the temporal dimension. Sonic fictions thus drag time into poetic ecologies of resonance, helping to mobilize the sense in which listening may perform, drawing in the critical vitality of spatial and temporal matters: a listening practice adept at understanding material facts as being connected to a mutational potential. Acoustic relationality is thus posed here not only as the basis for gaining awareness, or enlivening a sense of place, but equally for posing questions, for capturing other steps and grooves, to open a critical and creative view on what constitutes social reality.

Following these practices, relationality is cast as emergent—as shaped by intensities of transformation, or radical openness: Winderen's wild recordings, or Kubisch's electrical walks, force one into a world whose murmurings and energies are deeply foreign and yet extremely proximate; and Hoyt's musical and sonic fictions generate cultural mixes that suggest a mutational future already inhabited by our unhomed selves. From these particular experiences and sonic fictions it becomes possible to craft intersections of the proximate *and* the (other)worldly, the familiar *and* the unfamiliar, suggesting emergent modes of political subjectivity—dream-acts that figure themselves within and around the particularities of the social. Subsequently, I may ask: What particular impact might expressions of sound art have for thinking and approaching community, for approaching being-in-common? What consequences does such interweaving have for knowledge—What forms of acoustic knowledge may be grasped from acts of listening-dreaming? And is it possible to craft from the meshing of fact and fiction, work and dreams posed by acoustic relationality a greater sense of affinity as well as practices of care that extend toward other matters and lives?

Following this line of questioning, I want to reflect upon a project by the artist group e-Xplo. Developed for the Sharjah Biennial in 2007, *I Love To You* is a project based on engaging with migrant workers living in the United Arab Emirates. Through the project, e-Xplo was concerned to intervene within the complex reality of displacement, and the subsequent living and working experiences of migrant workers, searching for ways not only to highlight their precariousness or lack of rights, but to recast these toward a framework of agency, refusal, play, and public engagement. To do so, the group set out to record workers singing songs or reciting poems from their homelands; from parts of North Africa, as well as the Middle East, from Egypt to Tunisia to Saudi Arabia, a diversity of workers were recorded.[48]

For the subsequent Biennial exhibition, e-Xplo shared the recordings with local shops, offices, and restaurants, asking them to amplify the recordings over installed or existing audio speakers. Throughout the duration of the Biennial, the voices of migrant workers could be heard filtering through the life of the city with their melodies, languages, dialects, and songs.

e-Xplo's project works toward the crafting of an encounter, of bringing certain marginalized voices into the open precisely where relations are strained by abusive economies and state structures of inequality. Against and within such structures, these songs and voices are positioned so as to work at shifting the dominant tonalities of the streets, retuning the linguistic and vocal ecologies through the introduction and amplification of voices out of place, or strained by their particular emplacement. Installing or extending existing sound systems, these foreign melodies begin to intersect with existing aural patterns, riveting one to the other so as to reform the acoustic dynamics found in Sharjah.

Through the lyrical sonorities of song, expressed and shaped by the foreign voice, a voice out of place, or subjugated to forms of partitioning, the borders and demarcations of public life—of what is permissible and possible—are tensed or agitated. We might consider how the melodious dynamics of singing imparts particular meanings specifically where public statements and declarations are foreclosed, drawing out new delineations to the heard and the unheard. As Heimo Lattner, a member of e-Xplo, stated: "These voices in this particular place, under those circumstances, conveyed what cannot be said or addressed publicly."[49] These voices trespass the rather strict ordering of who speaks and where, encouraging, through the melodious volumes of song and itinerant qualities of recorded sound, a new social encounter.

In the project by e-Xplo, the musicality and lyrical expressivity found in singing are drawn out in negotiating experiences of place and displacement. Singing seems to allow for reaching into memory, resounding the *inner regions* to enable forms of recollection. Songs, especially, become carries of home and are unleashed in that instant of vocality—charged by the resonances and vibrations that tremble the chest, the mouth, the lips; a going out and a going in, this complex of returns that is resonance. One, in effect, returns to oneself by way of the resonant opening Nancy outlines, that primary vocality or cry which literally unfolds the body. As a potential act of recuperation—*bringing one back to oneself*—the resonant qualities of singing evoke a palpable link between oneself and a given situation, lending protection as well as guidance to those displaced, to those unhomed by the ongoing shipping of certain geopolitical orders. For the resonance figured by way of song builds a new volume, an emergent dimension constituted by the affective threading of oneself and the surroundings; a volume that may give way, through a type of repair and energetic holding, to a sense of coming home.

Here, acoustic relationality is a deeply tangible material—an "intermateriality" as Nina Sun Eidsheim suggests[50]—that assists in constituting the body, holding or suturing one together, returning one by way of a resounding vitality of sound as it moves in and around, and then back through one's body. This vocal movement, of unfolding and enfolding at the same time, of the deep breath needed to voice a new dimensionality, is further emphasized by e-Xplo in what they interestingly term the "inauthentic voice." As they highlight, their own desire to capture the voices of workers ran the risk of exoticizing the idea of an authentic voice, this authentically displaced, unequal voice. What emerges in the process of their work is rather an appreciation, even a pleasure found in destabilizing the notion of an authentic voice in favor of the inauthentic: this voice that may or may not come to represent the socially and politically oppressed subject.[51] That may, in fact, fictionalize.

While voice is often called upon to act authentically, to deliver the subject as wholly itself, it is also bound to a certain dissolution; as a sounded movement, voice both punctuates the body while diffusing beyond it. As Nancy poses, resonance is first and foremost a vocalization: a primary cry that fulfills the self by explicitly externalizing it, figuring a dimension that defines an inside and an outside, that defines space itself as space. In such resonant movements, voice comes to represent oneself by always extending beyond the body—to manifest not so much a disembodiment, but a "paradox" as Mladen Dolar suggests: the voice as both here and there, as being both myself and not myself.[52] To speak of an authentic voice then is to overlook the ways in which the voice is constituted by a complex movement, a state of dis/placement, figured as a paradoxical matter.

e-Xplo's work enacts a type of sonic fiction, however weak or weakly placed, tuned or out of tune. The diffusion of these voices within the streets of Sharjah may be heard to work across that particular acoustic dynamic I'm mapping: that of sound's dimension/less propensity. In following their work, I'm led to consider the ways in which an acoustic relationality, this double-listening may come to assist in negotiating the experiences of being far from home. Acoustic relationality, in trafficking between site and sitelessness, may figure as an important channel for working through the very conditions of displacement; it may allow one to speak about struggles over place, and social inequality, offering a track—a sonic fiction—by which to narrate the difficult state of statelessness, to resonate precisely where recognition is foreclosed.

The question of voice and experiences of displacement are additionally found in the work of artist Sarah Vanhee. In particular, her performance *I Screamed and I Screamed and I Screamed*, developed in 2013 at the state prison in Mechelen, Belgium, aims to draw attention to the hidden lives of prisoners. Through a series of proposed workshops, Vanhee set out to record the screams of those incarcerated with the overall ambition of creating a "scream choir." As with e-Xplo's work, Vanhee focuses on voice and vocalization as a means for critically navigating the politics of appearance and silencing, recognition and rights, which the institution of the prison draws out. As the artist suggests, "The prisoner himself does not have a face. His body exists outside the public sphere, it refers to an economy of illegality. His voice is inaudible."[53]

After a series of attempts to make her recordings, a process that led to a number of challenges as well as thwarted interactions with prisoners and prison staff, Vanhee finally presented a performance outside the prison walls. Standing in front of the prison, the artist prepared a script which she then "screamed" to a gathered crowd consisting of prison staff, police, and an art audience, aiming to give volume to the silence behind the walls.

> There is nothing more personal than a voice. Screaming is something very intimate. To open your mouth and let off sounds from a hole in the middle of your face. But you do it, you speak, or you scream, primarily, in order to be heard. And the idea of being heard, of possessing a voice nearly coincides with that of human and civil rights. Having a voice is much the same as having a vote.[54]

In their works, e-Xplo and Vanhee search for ways to reorder a dominant acoustic reality, performing to support the silenced or the dispossessed, the withdrawn and the marginalized. Through processes of recording and scripting, amplifying and screaming, their works shift from testimonial witnessing to acoustic performance, drawing out personal stories and life experiences through vocal performances and amplified recordings. In particular, singing and screaming are brought to bear upon the borders of appearance and recognition, specifically drawing one toward the stranger and the estranged. In this sense, their works attempt to constitute a public—a strange public—by carrying and caring for the voices of those who inhabit conditions of precarity and incarceration, utilizing the resonances of song and the rather shattering vocality of the scream as deeply politicized acts.

It is my sense that the artists do not so much speak for others as rework the distribution of the heard by shifting the borders of an audible and vocal ecology. As such, an acoustic relationality is posed, one that hovers over or interrupts an existing reality, to give way to a movement that may act upon experiences of displacement. In this regard, one may speak of an acoustics of repair and recuperation, which supports a *listening again*, a double-listening that moves from here to there, that hears by way of displacement. The amplification of songs from home, or the frustrated screams outside prison walls, wield an important intervention that come to tense, to vibrate, the borders shaping recognition and belonging. Screaming outside the prison walls, Vahnee literally calls upon a resonating force to incite a relation to those removed from society.

While Winderen and Grzinich engage with distant locations and ruinous matters, e-Xplo and Vanhee bring us closer to questions of voice and the politics of public life. Both tendencies contribute to the larger project I'm aiming for by highlighting the ways in which sound and listening, attunement and interruption may work to constitute modes of social engagement that extend or *move* the very notion of sociality. Following the dynamics of poetic ecologies of resonance, their works remind how public spheres, and the discourses often defining political life, are not always recognizable as such. Rather, their works are suggestive for considering the "stranger sociability," outlined by Michael Warner, as constituting publics and counterpublics alike. As Warner argues, the public sphere, founded on the articulation of public discourses and a community of values, is radically bound to a "social imaginary." "A public is always in excess of its known social basis. It must be more than a list of one's friends. It must include strangers."[55] Publics and counterpublics are thus supported through a deeply potent imaginary, which enables groups of strangers to essentially understand themselves as being bound to a social community, as being-in-common.[56] In this regard, stranger sociability must be appreciated as underpinning the public sphere and its discourses, granting a power of circulation as well as connection.

If, as Warner suggests, the public sphere is figured through a social imaginary, publics are also prone to fracture, destabilization, reinvention, even poetics—what Warner additionally identifies as the "poetic world-making" tendencies of public discourse.[57] While publics may imagine themselves as sharing in forms of speech and hearing, discourse and its circulation, and that remain bound to stable identities and meanings, to the narratives of national identities Gilroy seeks to complicate for example, it is clear that such stability is partly a fiction. A fiction that forms of public art making often seek to unpack, mobilizing the embedded strangeness and poetic or fictional undercurrent always already constituting public life. As e-Xplo and Vanhee demonstrate, there is often more to the public sphere than meets the eye.

The methods and acts often brought into play in works of sound art can be appreciated as accentuating public life as one of strangeness—that strangeness, and the notion of the stranger, in fact runs throughout the imagined stability of publics, even counterpublics. This is greatly extended if we consider the practices of Winderen and Grzinich, whose works specifically integrate the nonhuman and the world of matter, and therefore introduce a "planetary" perspective in terms of what we may "socialize" with. In doing so, I'm interested in what becomes of recognition, intelligibility, and even mutuality when we enter a "public sphere" replete with this that I can never know or recognize, but which I may hear or feel nonetheless. A strangeness that might be said to always hover within the continuum of sound and the affective resonance of mutuality: a strangeness that continuously displaces place, others oneself, detunes attunement, inauthenticates authenticity, troubles harmony, overexcites speech.

Poetic ecologies of resonance are constituted by relational intensities and movements that force new alignments, that reorients one precisely along the complicated borders to what one may never understand: the foreign languages that suddenly transform certain streets, and the screaming that attempts to carry so many unseen identities beyond their capture; or the strange murmurings of marine life that exist across the planet. Poetic ecologies of resonance, rather, foster a fruitful expansion and fragmentation of sense and nonsense, figuring a dream-activism, an extra-sense—*I am always already somewhere else, tuned to some-thing or some-one that touches me with its vibration, its resounding intrusion, its energetic transmission, and that figures one as a witness, a partner, a co-actant within its migrations*. Accordingly, acoustic relations engender complex alignments and associations, rhythms and repairs, which, when oriented toward social organization and public assembly—situations fundamentally based around speaking and hearing, sounding and attuning—produces a mutational form of sociality, that brings together the familiar and the unfamiliar, the authentic and the inauthentic, matter and energy, into forms of expression.

Audible Dwelling: Shelter/Amplify

I've been considering sound art as a complex relational practice that opens toward expressions of extra-relationality. As such, it allows for focusing on questions of the situated and the factual by relating these to encounters with the foreign and figures of strangeness—toward the limits of recognition. Consequently, works of sound art often approach the border, the social circle, or the foreign territory not so much as a barrier, but rather as a surface of vibration, an animating conductor that supports passage across interiority and exteriority, the seen and the unseen, which further drives a negotiation with the powers and histories of exclusion.

In this regard, works of sound art often search for ways to generate interactions between the familiar and the unfamiliar, the host and the guest, the told with the untold, putting each into a compositional structure, a poetic ecology of resonance by which to support listening as a work of reorientation—as a passage. Phenomenological fact and sonic transformation, the immediate and the imagined, these emerge less as separate territories and more as an intensity of relations—as routes or steps by which to figure other alignments and constellations of affinity. Whether in the form of an improvised touch, a musicality of signals that detunes the material state of things, that sets one dreaming upon the objects of the world, or by pirating the resonances and rhythms passing across the black Atlantic, or in the amplification of the voices of those held behind bars or captured by the infrastructures of an uneven geopolitics, acoustic relationality is posed here as the broadening of listening toward the unheard as well as the not-yet: an ethnography precisely around that which no longer *keeps me at a distance and that exceeds description, textuality*. Here, what constitutes one's "situatedness" must be seen to relate equally to that which exits the situation, that beats out of time, withdraws from narration, and estranges my sense for what the situation is: a fugitivity, a sonic fiction. Within poetic ecologies of resonance, the unrecognizable raises the stakes on performing situatedness, explicitly amplifying the need for other methods, other ways of recording, and other ways of listening.

The acoustic position I'm mapping is therefore aimed at underscoring listening as an affective labor, as a gesture that may critically and creatively upset the order of hearing so as to give way to the previously unheard: not as an act of capture or recognition per se, of making legible that which is illegible; rather, to *listen with*: a double-listening that may attend to the here *and* there of modern, global life. This is not to overlook the deeply challenging realities of racial struggle, and the degree to which globalization demands an ongoing fight against systems of violence and the making of disposable subjects. What I'm aiming for is to draw out pathways that may assist in such struggles, to wield a certain criticality in order to enable other social and political imaginaries; to find other ways of recounting violence as Saidiya Hartman argues.

Listening, as I'm proposing, stays unsettled, or, as Evelyne Glennie suggests, it is bound to an "endless curiosity."[58] In this sense, listening is adept at contending with and relating to the limits of the recognizable—it can do much in struggles over recognition by explicitly moving the lines by which recognition takes shape, is legally held even. As Achille Mbembe argues, within our current state of globalized life one must have the "courage to listen," to bring one's listening forward to confront the previously unheard—that which pulls at my body, and the delineations of the home, those who cross the borders, for example, to arrive at the door.[59] This act of courage might be found by nurturing a sense for *being-open*, for taking the risk listening requires as that which opens onto the world and the world onto oneself, and through which resonances may be tracked—a tracking, as Satch Hoyt suggests, that traces and attends to the ceaseless migrations of cultures and of people. As a consequence, such listening, such openness may be followed as a way of taking up an endless journey, to follow the poetics Moten poses as that which never settles, that is social life itself. Tracking and retracking the very tracks that have been placed before us, or those emerging within current migrations, and whose remixing is posed here as the enactment of an acoustic relationality, a poetic ecology of migrating signifiers whose rhythms and resonances may support engaging with the held down, the historically contentious, as well as the stranger at the door.

Subsequently, I may further ask: Is it possible to evoke an acoustic scene, a remix, an improvisation that assembles the nurturing touch of an intimate life with the strangeness always already nearby? One that constructs a resonant body—a *resonant assembly*—out of the drives of compassion, *co-being*, and the wayward interruptions of sonic matters? Finally, can we think of sound art as an art of taking care, forming a new type of shelter constituted by what Trinh T. Minh-ha terms "elsewhere, within here"—the restlessness underpinning the poetic crafting and recrafting of strange alliances?[60]

* * *

I have been touching upon a range of works of sound art, works that use sound and listening as operations, as well as a framework by which to approach the unseen and the foreign, the distant and the unintelligible. In this sense, I understand sound art as an experimental set of practices and methods driven by a desire to hear beyond the hearable, to touch the flourishing of life beyond the human and the textually bounded, to once again track (or scratch) the tracks that have been laid before precisely because we know there is more to find. In this sense, sound art contributes to the concept of acoustic justice by extending the right to listen toward a broader notion of sociality and which may impact onto a politics of recognition. Following sound art, acoustic justice emerges as a sonic fiction itself, especially attuned to the particular qualities and articulations of dis/placement, of otherworldliness, and the fractal intensity of the mix.

As a final track, a further instant of listening-dreaming, I want to consider a work by the artist team Learning Site. In particular, their project, *Audible Dwelling*, navigates the complexity of the personal and the public, allowing for a consideration of how acoustics supports a transformative process. In this case, the work does so by explicitly drawing into the open a performative architecture, an act of "sheltering" that seeks to unsettle or animate the lines of a public sphere.

Audible Dwelling is an architectural work that occupies public environments and addresses issues of housing and urban rights. It consists of two structures placed side by side, forming a stereophonic partnership, each of which is divided into two rooms or sections. The first section functions as a small bedroom or office that integrates audio recording and playback equipment, while the second section works as a more open living room. Importantly, the first section also functions as a speaker enclosure operating to assist in the transmission of sounds through a set of audio speakers mounted on the exterior of the structure. In this regard, *Audible Dwelling* is an architectural structure that also works as a sound system; the design of the structure serves to both facilitate the amplification of sounds to a broader public as well as cater to a social process that has at its center the question of housing in the city.

In addition to the overall architectural design, Learning Site decided to construct the interior of *Audible Dwelling* in reference to the modernist design

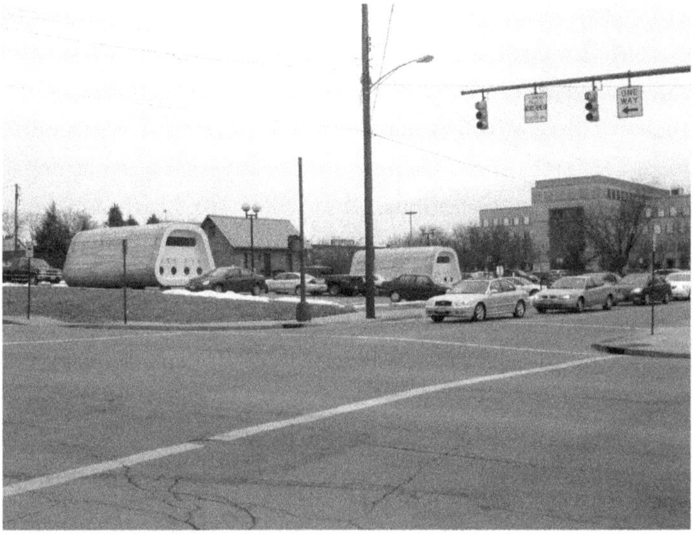

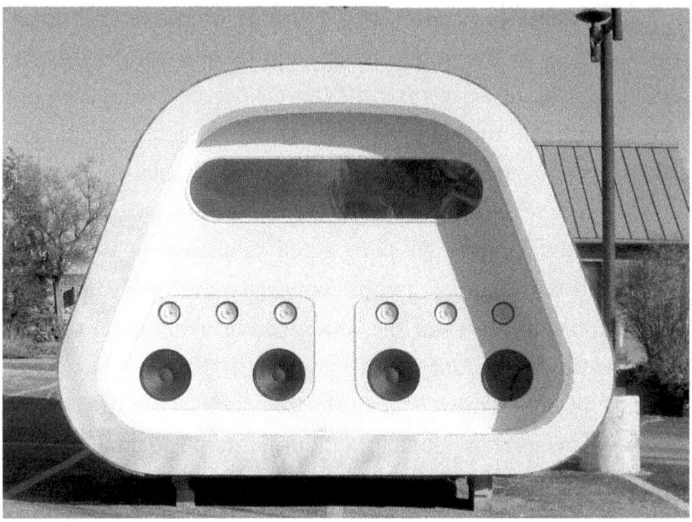

Figures 2.1 and 2.2 Learning Site, *Audible Dwelling*, Columbus, Ohio, 2009. As part of the exhibition *Descent to Revolution*, produced by Bureau for Open Culture. Photo: courtesy of the artists.

work of De Stijl, which was suggestive for how design may contribute to debates around housing. In particular, Eileen Gray's *De Stijl Table* (1922) was chosen as a main design reference, and was used as a formal guide for facilitating the experiences of working or living inside the structure as well as the amplification of sound based on "transmission line" speaker construction—a form of speaker technology that works to harness rather than absorb the resonances produced by speaker enclosures. In this way, Gray's *Table* was architecturally elaborated into a larger interior structure acting as a desk, a shelf, a bench and a bed, and also serving to acoustically support the amplification of a sound signal. *Audible Dwelling* behaves then as an extended resonant chamber: a structure putting into motion an acoustic, sonic gesture aimed at supporting social engagement.

For its presentation in Columbus, Ohio (2009–10), Learning Site responded to particular social histories related to the given site of its installation and its connection to immigrant communities, as well as questions of private and public urban development.[61] Recording residents of the city reading texts, recalling memories, and arguing against new private developments implemented by the city, as well as playing back scripted and staged speeches, the work performs across both public and private registers. As such, it aims at mediating between a personal and public imaginary, while at the same time discarding the maintenance of these as separate; rather, the work gives articulation, through a form of resonating amplification, to a plurality of issues and questions that tenses understandings of the public sphere and its rational discourses. In other words, it delivers a dream-activism.

For a period of six months, *Audible Dwelling* acted as a generator of activity, capturing entrenched views and concerns circulating within the local neighborhood and lending to public debates as to the future city. In this sense, I'm interested in how *Audible Dwelling* acted as a form of public megaphone giving traction to a range of public concerns and personal experiences. Interestingly, the amplification of the thoughts and feelings of residents found support by relocating the act of people standing in public space to voice their opinions to that of an architectural performance; as Learning Site proposed, *Audible Dwelling* became a "speaking house"—a performative architectural device enabling speech to find another relation to its public voicing. In this way, *Audible Dwelling* both sheltered *and* amplified the expression of diverse views.

Through *Audible Dwelling*, Learning Site raises questions as to the nature of public demonstration and political speech: What form of speech is this, and in what ways are we trained to listen to the voice of politics or resistance? And further, what sense might one have for being able to speak out publicly, and in what ways might one be listened to? As an intervention, *Audible Dwelling* raises questions about how speech is situated within particular contexts and communities, and how one speaks through those contexts and communities, navigating a range of acoustic, institutional, and social structures. By amplifying local voices aimed at the public sphere, *Audible Dwelling* does not simply playback voices; it additionally shelters them, situating questions about housing, real estate development, local history, and labor politics within a strange dwelling, one that grants movement to new social expressions.

Audible Dwelling may, along such lines of thinking, be imaged as a ship, a strange vessel whose resonances are cast in order to make of the city an open sea. By placing the one who may stand up in public within a more secluded space, *Audible Dwelling* inaugurates public debate in such a way that might reorder the normative lines defining the ways in which such debate may occur, and, further, question what types of languages and speeches might resound within such constructs. In this sense, as a vessel *Audible Dwelling* works at opening the arena of public life, resonating its acoustic frame so as to instigate a greater sweep of audibility and imagination.

Through its performance, *Audible Dwelling* unfolds a process of creative habitation that fosters reflections on the reality of coexistence within the city of Columbus; it steers us toward specific realities while searching for a transformative reworking, drawing out local memories and losses, fantasies and fictions, so as to reconstitute public engagement. In this way, *Audible Dwelling* draws from sound a capacity to both speak toward the fact of city life while, at the same time, moving such facts toward horizons of social transformation. As such, it asks of us to literally dwell upon the often tensed lines of public life, occupying and inhabiting contested sites and situations in such a way as to support a working through of how public life is full of personal lives and dreams. That sound and listening provide a means for enlivening the narratives and social imaginary of city life leads to a greater understanding of the potentiality of an acoustic act, as an art of giving shelter to often emergent (or repressed) forms of collective life.

Listening as Extra-Sense, Resonance as Refuge

I have been considering different works of sound art in order to underscore the ways in which acoustics can be appreciated as a relationality—the establishing of a relational frame through which sound and listening are worked through as transformative and transportive acts or experiences. This has led to mobilizing the figure of resonance, and the notion of poetic ecologies, in order to better engage the relational complexity and potentialities of acoustic matters and movements. From Jana Winderen to John Grzinich, along with Christina Kubisch, I question the ways in which sound art brings us close to the foreign, the distant, and the unfamiliar, from nonhuman species and worlds to the untold histories and electrical energies of place, drawing these out through improvised contact and fieldwork. Such acts do more than discover the nature of other life; rather, they intend toward a relational intensity: a scene and sense of involvement, a transformative closeness, a listening. In what ways do such encounters and attunements, such involvement with the nonhuman impact onto forms of recognition, in terms of a desire to hear beyond the intelligible? If recognition, as Axel Honneth argues, is central to fostering a "moral grammar" of the social, in what ways do understandings of community alter when hearing the nonhuman, as that which is unrecognizable? In short, does recognition require a certain legibility and, therefore, a certain ordering principle that presumes what being-in-common might be? The sonic fictions of the black Atlantic offer a useful and challenging framework for furthering discussions on acoustic relationality. By working at negotiating the complex intersections of place and displacement, past and future, as shaped by black Atlantic passage, sonic fictions offer routes for navigating and narrating the mix that is modernity—a brutal, violent, resistant, and funky mix shaped by the double consciousness of diaspora cultures. The strange compositions and transcultural mixes that shape sonic fictions give over to other forms of listening, other ways of dealing with relations, other tempos for contending with power, of staying on the move. Throughout such discussions and reflections, poetics and resonance, and their various grooves, intertwine, and are understood to underpin sonic fictions, the sound arts, energizing the extra-relational intensities listening seems to afford. Listening-dreaming is posed then as a form of navigation—a way of orienting upon the troubled waters of the mix, and the strange compositions emerging by way of poetic ecologies of resonance: for working at the transformative or mutational logic of acoustic propagations. Through the work of e-Xplo and Sarah Vanhee, the question of voice is brought forward. What becomes of voice from within the framework of resonance, and the extra-relational prompted by its excess and emergence? The voice is mobilized as the primary act of resonance itself, that first sound which, as Jean-Luc Nancy suggests, evokes the self to itself. As such, voice resounds precisely as *being-with*, suggesting that while the self may emerge it does so always already estranged and strained by an exteriority. The voice, in this regard, also enables negotiating and navigating that complex, external world. By reflecting upon the project *Audible Dwelling*, poetic ecologies of resonance are figured as enabling forms of social engagement and public demonstration. In this instance, questions of urban struggles are engaged by reorienting a sense for the political voice, one that might find traction by occupying the public sphere through other means. *Audible Dwelling* articulates the ways in which sound propagates to traffic across the facticity of real life, the spectral, and the mutational, lending material force to the imaginaries underpinning public life.

Following the methods and matters posed by forms of sound art, I'm led to an understanding of acoustic relationality as what may enable an approach to the intensities of social and global life, specifically forcing understanding as to the ways in which the familiar and the unfamiliar, neighbors and strangers, human and nonhuman coexist in tension, as an ecology of resonance; for is not resonance fundamentally a connectivity held in place by that "dynamic event" Hartmut Rosa describes, and its propensity for change and transformation? Sound art, in this sense, can be appreciated as explicitly attuning one not only to all that resides here and now, within the social or architectural field of meaning and legibility, but also for spiriting the interruptive force of noise, a resonance gone too far, a poetic extra-sense, which sets the conditions for greater affinity with others—with the other that resides on the limits of the familiar.

In this regard, questions of recognition and justice are extended by challenging in what sense these terms may carry the complex relational dynamics of "social reality." From the planetary world we inhabit to the fractal identities and strange public spheres constituting contemporary life, in what ways might social recognition take place? And how might understandings of just relations be found, following the idea of poetic ecology, which seems to require moving social justice toward forms of Earth justice? Following such questions, it is my sense that approaches toward working at recognition, or just relations, demand a greater mobilization—a radical imagination.

Unseen life-forms and electrical energies, songs and screams that produce strange publics, and the amplifications that work at linking and elaborating cities, to mobilize local possibilities, oceanic possibilities—such gestures and poetics point toward a central thread within acoustic relationality: the capacity to *listen both ways*, a double-listening adept at drawing out a range of sonic fictions. Listening in this way, as I've tried to suggest, nurtures forms of dream-action that may also work at repair, where "listening is always already healing."[62] From experiences of displacement to alienation and dispossession, listening is often mobilized specifically as a means for contending with difficult memories and histories of loss; a listening that may enable a threading of the told with the untold, a making room for telling that which is never quite tellable. Listening may come forward as an extra-sense by which to hear beyond a particular order of audibility, and to form solidarities with unlikely partners, from creatures to signals. Acoustic relations become important spatial and temporal forms, the building of other dimensions, as well as archives and shelters, where one may find refuge in a particular rhythm or tonality.

Nathalie Harb, an artist working in Beirut, underscores this potentiality within listening, and the ways in which modulations of the distribution of the heard work at repair and the reconfiguration of experiences of real life. Her public installation, *The Silent Room*, presented in 2017–18 in Beirut, is a form of shelter located on a busy street in the city. Consisting of a partially closed wooden room mounted above ground-level and accessed through stairs, *The Silent Room* becomes a soft enclosure within an otherwise open and crowded environment. Visitors enter and lie on the floor, letting the sounds from outside pass over their bodies, their thoughts. The architectural shelter performs an extremely subtle yet effective filtering process, reducing the intensity of external input; immersed in shadow, lying together with others, one retreats from that sense of exposure experienced while on the sidewalk, with the traffic rushing by. Instead, within *The Silent Room* visitors retune their sense for the ongoing noises surrounding them; one might say, visitors find a certain balance through this time and space of listening, a *listening again* to what is so familiar. What emerges, instead, is a flow of sounds—a resonant continuum that, rather than impair one's senses, supports relaxation, recuperation, *escape*. In *The Silent Room* traffic noise becomes an oceanic flow.[63]

The experiences of repair and recuperation take on greater urgency if we consider the social uprisings, conflicts, and general state of catastrophe that have swept the globe in recent years in the wake of financial, democratic, ecological, and public crises. The level of charged debate, political battling, social movement, and the heated narratives that thread themselves around issues of public life and private loss, and that increasingly appear through social projects, new forms of commoning, self-organization, and alternative pedagogies, along with forms of illiberalism and new nationalisms, all are indications of how pressing questions of togetherness are. The emerging practices of more direct modes of governance and solidarity movements are a counter-measure, within an American and European context at least, to the dismantling of state welfare and the eviction of public culture, and the crises that face many today. Such counter-measures come to actively fill the gap left behind with intensely creative and nurturing approaches toward social and ecological worlds, and which are increasingly called for: caring for planetary life, working at intersectional social solidarities, extending compassion and recognition beyond the immediate and the visible, radically shifting the positionality of neighbors and strangers, hosts and guests—giving way to new openings for cooperative practices and a rethinking of democratic politics, from deep commoning to radical exit. Within such worldly and planetary mindfulness, voicing and listening, sharing and finding place, and constituting social formations explicitly with those out of place, held down or beyond recognition, take on pronounced urgency.

Acoustic relationality, the works and dreams of sound art, and the fostering of poetic ecologies of resonance, I hope may point toward another sense for what constitutes the facticity of things and the operations of real life. These acoustic attributes and constructs rather locate one within a framework, a sonic fiction, defined by stranger relations, a listening-dreaming that may set to work at reorienting (or disorienting) the intelligible by relating one explicitly to what cannot be related to: this other species, this metallic object, this complex history, this stranger whose language one may not fathom and that may require more than one listening. The work of listening may be underscored then not only as an experience of hearing, but equally as an act of engagement, even negotiation and certainly joy, of care and celebration, which figures one alongside the complexity of the unrecognizable. Here, acoustic justice turns us toward the planetary position that Dipesh Chakrabarty outlines as one that reminds of our human "creatureliness"—this creatureliness that is mutational by nature.[64] And in doing so, forces us to recognize the deep time and deep movements to which we are but a particle.

3

Skin-Work

Queer Acoustics, Borderspaces, Economies of Desire

An affective and corporeal intensity underpins the experiences of sound and listening; the noises and resonances that animate the matters of the world to galvanize and enrage, to coalesce and congeal, to nurture and thrill—to press us, and to pressure one with another: *this other that softly wraps or that pierces with its soundings, its awkward silences*. Such intensities leave their traces and echoes.

The movements of sound are deeply pervasive, driving from all directions a force, an array of voices and vibrations giving charge to all types of bodies and surfaces, and trafficking across interiorities and exteriorities, physical matters and psychological recesses. We are susceptible to its stirrings, these sonorous rustlings and punctuations, and as such, we stagger in their wake and celebrate the drive they incite. The movements of sound figure the body as an erotic sensitivity, a "membrane" as Steven Connor suggests,[1] one tensed by echoes, vibrations, and rhythms, sympathies and agitations—*to shiver with sound*. In this way, sound and listening are fundamentally a question of touch.

The experiences or capacities afforded by way of sound and listening bring one into a mode of relationality that supports as well as exceeds the parameters of speaking and being heard; rather, the flows and ruptures of sound continually press the borders of the dialogical, punctuating speech with amplifications, murmurings, and interruptions. As such, aurality is often never only a question of listening to another, but rather relates itself to a broader tactile, sensual, and psychological weave. This comes to support capacities for joining together, figuring a social and material means of connecting, as well as what may cut the ties binding one to another, scarring the (psychic) flesh with noises and the definite rush of social forces.

Considering these affective, tactile, and psychological intensities of aural experience, the skin surfaces as deeply impressionable, as fundamentally participant within the acoustic figuring of self and other, subjectivity and the relationships defining it. I want to move toward questions of touch, not only to engage acoustics as a broader spatial, social, or ecological operation but also to draw back, closer to the body. For if, as I'm posing, acoustics underscores the potentiality for working at orientation, offering a means by which to navigate the relational and extra-relational coordinates that figure one within a range of experiences and struggles, questions of sense and sensuality are equally drawn out. Acoustics may give definition to a broader arena of relationships, but it does so by also affecting the inner touch of sensation, intimacy, and the experiences we have as sensitive bodies.

Moving from questions of touch and tactility, I want to follow a number of theoretical perspectives, drawing out the skin as a complex border. This will further allow for questioning other forms or practices of bordering, as found, for instance, in the new border regimes of the globalized world. By questioning the ways in which sound, listening, and acoustics draw into play tensions around borders, from the bodily to the national and transnational, I aim to follow the skin as a site of relational and political work.

The Skin-Ego: Impressions

In his work on the question of subject formation, the psychoanalyst Didier Anzieu poses the concept of the "skin-ego," reflecting upon the skin as a complex surface, a membrane, a parchment that shapes one's (psychic) body. The skin is positioned through Anzieu's theories as a highly sensitive lining, one susceptible to the familial, social, and material conditions that surround and mark us with their force and bonds, highlighting the skin as the means by which subjectivity emerges, is drawn out, and continually works through experiences of being situated. Accordingly, the skin-ego acts as a deeply generative arena of envelopes and sacs, punctures and caresses that provocatively align the psyche with the flesh.

> The first function of the skin is to be the sac that contains and retains inside itself all the good, full material that has accumulated through breast-feeding, everyday care, and the experience of being bathed in words. Its second function is to be the interface that marks the border with the external world, which it keeps on the outside, the barrier that protects one against being penetrated by the aggression and greed of others, whether people or objects. The third

> function of the skin, which it shares with the mouth and carries out at least as much as the mouth does, is to be a site and primary mode of communication with other people, to establish meaningful relations; in addition, it is a surface for registering the traces left by those others.[2]

By way of a series of functions, the skin emerges as a dramatic and extensive mechanism of mediation between interiority and exteriority, facilitating protection and self-preservation while negotiating the intrusiveness of the outside. Skin fundamentally *enwraps*, holding in the stuff of the body with its supple lining—this sac that grants one definition. As such, skin is what confronts the world around; it meets the world, to assist in navigating the material and social shape of things, and how those things are made available or not, put in reach of oneself or withdrawn.

The skin is also a surface stretched amid other surfaces and, as such, it is susceptible to the pressures and textures of contact—it may be caught off-guard, suddenly, under the force of a certain object or action, scarred by the weight of the world, or the greed of others; or it may loosen, to give way to the sensuousness of pleasure and intimacy. Under the force of social contact, the skin shivers and puckers, it ages, collects time in its folds and creases, defining the particularity of myself—this skin that is weathered as it lives. It is the literal surface whose colorings and textures, folds and protrusions, pocks and slacked corners, presents one as an individual, *this shape and surface I am*. Accordingly, it may, in fact, be what I give to another, this skin and all that it carries, signifies, performs, in the touch of so many caresses. Skin is what I can offer, in the movement of myself, this movement which ultimately brings the skin to life, drawing it up close, or retreating, animating this surface that I am.

As Anzieu highlights through a range of theoretical perspectives, the skin comes to define one's first contacts, giving way to a life of exchanges and interactions that leave their marks, that impress onto the body and oneself the familial and societal frames one works to inhabit: in moving into the world, the world gives way, or it resists; it pulls back to accept this body, receding in support of the many wishes and desires, or it hinders passage, forcing one to retreat, to recoil, sending the skin shivering with fear, anxiety, and the recognition of one's limitations. As such, the skin receives a deep education on the nature of the world around, and the nature of oneself as a body meeting that world.

The skin is therefore a fundamental layer to the material, social, and psychological experiences of being a body, a subject, formed as the membranous site around which one's "bodily figuring" unfolds, is literally defined. Skin emerges, following Anzieu, as the site by which one receives and gives, acting as an extremely sensitive point of contact, and the movements between interior and exterior, and importantly, attachment and detachment. As such, it is poised as a connective matter wrapping itself around not only those one loves or are loved by, but also the spaces of home, neighborhood, work, and all the institutional situations into which one is placed, extending what Merleau-Ponty terms "the flesh of things."[3] Through such movements, such "inter-sensorial" connecting,[4] which I may image as a *throwing or pulling of one's skin*, one works at finding place, to create place for oneself, and from which all the subsequent movements of social accommodation or resistance are prompted. Through such movements, the skin disciplines the body on the properly behaved and permissible. It educates not only on the inter-sensorial and inter-material contours of oneself in the world, as being in particular places and how those places extend or cut into oneself, but additionally, it provides an ongoing performative contact with the codes of conduct shaping bodily and social orientation—a "consensuality" positioning corporeality as a question of the common good: how one is pulled into or pushed out from forms of community, of being-in-common, of consensus: a sensing that is equally and immediately operative, political.[5] Through all types of deeply vibrational, tactile, and inter-sensorial experiences, one learns of the connective threads binding one to others, and how such binds are tensed, played, or broken. *I am grounded in my skin, and it is through my skin that I find my ground.*

Queer Acoustics: Orientation and Affordance

In considering the skin, I'm interested to draw out experiences of sound and listening, as well as the question of acoustics, as what supports relating one to another, in the exterior spaces of the social—the extra-relationalities that put one into contact with what is *out there*; in addition, I also want to move closer, or inward, to questions of embodiment, of bodiliness, this sensual figure I am. Here I may emphasize an acoustics that specifically facilitates deeply psychological and sensual experiences, the processual or metabolic dynamics of bodily life; an intimate acoustics enabling the sharing of skins, the whispered words of intimacy, of being close, closer, and the trembling sensitivities defining touch, of touching and being touched. Such considerations allow for understanding acoustic justice as a question not only of social recognition but also of bodily orientation, of sensation, and how one may work at particular forms of togetherness through the tactile stirrings underpinning what it means to hear. For hearing, as Connor reminds, is a form of touch.

Yet, it's important to emphasize that skin is never just any skin; rather, skin is always a question of the particularities of cultural identities, underscoring the body as situated and aligned. In her book *Queer Phenomenology*, Sara Ahmed challenges the ways in which traditions of phenomenology, in particular through the work of Husserl, often overlook the more socialized, racialized, and gendered shape and impress of the phenomenal; the objects and things, the architectures and rooms that surround are never truly neutral, never only there materially, phenomenally, accessible to oneself, but rather are *made* available through a range of highly situated, historical, economic, and social processes and precedents that work to establish the normative shape of what we may associate with and how.[6] For Ahmed, one's bodily figuring in the world is thus always already defined by a set of dominant constructs and genealogies that are deeply material and spatial, and that enable or constrain the particular grasp specific bodies may have on the world around. As Adrienne Rich notes, "To locate myself in my body means more than understanding what it has meant to me to have a vulva and clitoris and uterus and breasts. It means recognizing this white skin, the places it has taken me, the places it has not let me go."[7] One thereby gains entry or not according to the availability of passages and pathways, and how these open for some more than others. In short, bodies are never only just bodies, but are already shaped by social and political norms, by history and the "politics of location" which reads one's skin in defining the phenomenal availability of things—their perceptual and physical availability according to the social, racial, and gendered specificity bodies and spaces carry.

> The lines that allow us to find our way, those that are "in front" of us, also make certain things, and not others, available. What is available is what might reside as a point on this line. When we follow specific lines, some things become reachable and others remain, or even become, out of reach. Such exclusions—the constitution of a field of unreachable objects—are the indirect consequences of following lines that are before us: we do not have to consciously exclude those things that are not "on line." The direction we take excludes things for us, before we even get there.[8]

Following this questioning, Ahmed opines an important view of how "orientation" is never freely found, but rather is shaped by the established patterns and processes that bring one into certain alignments, or that make particular misalignments possible as well as dangerous. One is equally oriented by the world as one makes orientation for oneself; one follows certain lines, which consequently exclude or obscure others. To orient is thus to be situated, to find one's bearings, within a given space as well as within or against particular social structures and systems; to be figured as part of the processual unfolding of one's bodiliness. Orientation is equally a performative operation, whereby one may seek out particular support through the material world while contending with the lack of availability or not of certain things: one is often pushed onto certain tracks, or forced off others. One therefore *practices* orientation, is situated by it, which shifts as bodies shift, as one aligns or misaligns, attunes or disturbs, is welcomed or pushed out.

This line of questioning leads Ahmed to the concept of "queer phenomenology," which problematizes the seemingly smooth matters of worldly contact and asks for greater recognition as to how our actions are oriented by way of particular lines, and further, how in moving toward things we also gravitate around certain alignments and synchronizations: in following specific orientations we also support what they come to represent. In this sense, Ahmed puts phenomenology to work by considering how orientation is often a form of "lining up"—a "falling in line" derived by way of a dominant heterosexual ordering, where "being straight" is often to "straighten up." For orientation is already in place, is defining of place, before one steps in; bodies are already figured, situated, framed, and oriented in certain ways, in line with particular directions, particular architectonics and genealogies, existing orientations. To be a body is to already be placed; to be called into order, and to find one's "standing" within a given scene or community. To fall in line thus occurs on many levels, conscious and unconscious, bodily and linguistic, social and familial, and has consequences for how one moves through the world, through particular contexts, making possible or not the articulation of certain forms of life. In response, Ahmed seeks to veer off-course, to pose queer orientations, which act to "put within reach bodies that have been made unreachable by the lines of conventional genealogy." As such, "queer orientations might be those that don't line up, which by seeing the world 'slantwise' allow other objects to come into view."[9] Queer phenomenology thus poses the very problem of orientation, questioning in what ways it sets the terms by which "deviance" is cast.

I'm interested in following Ahmed, and what she emphasizes as "the work of reorientation," as what assists in establishing other lines—that puts orientation into question. In doing so, I'm concerned to additionally *queer* the acoustic, giving accent to the ways in which acoustics and acoustic practices assist in processes of (re)orientation that additionally negotiate the dominant tonality of a place or location. Voices find resonance within environments according to the availability of particular acoustic matters—those who listen, who support the articulation and movement of voice, as well as those things that invite one to speak or not, the material forms or social arrangements that welcome or that care for certain bodies and their sounds, their articulations and reverberant meanings; acoustic arenas whose configurations enable forms of attunement or disagreement.

A politics of location, and the corresponding reading of skin, as Rich suggests, lends to the rhythms of place-making, of finding place, which is always a question of orientation, a rhythmicality that lends definition to how and where we may move, a moving away or a moving toward, dramatically enabled and enhanced by the material and social supports around us—by the spacings and timings that open or close. Rhythms may work at demanding entry, bending or breaking the organization of a particular scene or situation so as to move differently, to give expression to an altogether different pattern, a *veering off-course*—to refigure the social or cultural axis upon which movements occur. Such work of reorientation is equally articulated in the reverberant gestures one may enact, the ways in which we echo-locate in navigating place, sounding out possible routes through all that may respond and echo in return.

Acoustics is about the material or social supports that enable movements and articulations, supporting one's orientation through the propagation and reflection of certain sounds and their meanings. In this way, acoustics is always already informed and effected by the normative patterns or schema that often define spaces, their formation and function, contributing to what can be heard and where, who may speak or not, what types of behavior can appear and by whom, what comes to be understood as "deviant" sound, and to the ways in which we may enact forms of togetherness in support of the flourishing of a certain vitality: the sharing of speech and skin in order to work at particular forms of orientation and organization.

Following Ahmed, a queer acoustics may pose an interruption—a distortion, a coloring—onto the particular tonal shape of a place, detouring fidelity as what aligns one's listening with a particular acoustic norm or dominant sonic image. In doing so, a queer acoustic may allow for other resonant flows and connections, other configurations of sounding and listening, other meaningful synchronizations, to rework how one is oriented: to lend to what it might mean to veer off-course, to listen slantwise. Queering the acoustic may enable the retuning of a sonic horizon, surprising our auditory world with the rarely heard, or with an altogether different reverberation: these gestures, these interruptions, these resonances or echoes that disturb or mimic, amplify or elaborate the distribution of the heard. It may also work at shifting one's body, and the deeply impressionable surface of the skin as it is touched by the movements of sound, by the vibration of the flesh of things, as it seeks to touch that which is held at a distance, or that which is off-limits.

In posing acoustics as a fundamental shaping of sound, and by extension the alignments and coordinates of one's situatedness, I'm moving from notions of skin, as what contributes to articulating one's bodily definition, to the ways in which one may work at reorientation: to find the means for figuring oneself differently. This leads to underscoring acoustics, and *acousticking*—the articulation of acoustic acts—as what may enliven and thicken the work of reorientation and its impact on the order of things. From an acoustics of intimacy, the whispered words extending speech and hearing as what touches, to the movements that seek other routes, differentiated by way of critical and creative echoes and rhythms, acousticking pulls into place an array of supports as one finds one's way. Acoustics, as I'm underscoring it, is implicated within the work of reorientation, a form of work which I want to consider by way of the term "skin-work," specifically giving definition to the sensations I may feel and act upon in a certain room, a certain neighborhood, this social life in which one is placed. Skin-work thus works along the lines connecting the body to a greater construct of material and social lines—an architecture of lines whose ordering and reordering become a question of sensuality, consensuality, consensus: a work aimed at orienting and reorienting forms of "common sense."

Anzieu suggests as much through his theory of the skin-ego, and what he further defines by way of the "sonorous envelope."[10] The notion of the sonorous envelope is developed as one of a number of modalities of skin-work at play in subject formation; in this case, the sonorous envelope underscores the experiences of sound as a vibratory, immersive flux that surrounds and bathes the infant in an ongoing flow and, at times, flood of energy. Skin is activated and brought into play by the particular energies that sound delivers; this bath of sounds and vibrations, echoes and tonalities that surround the body, and whose "fluid" qualities create an especially nurturing as well as disturbing contact with the world: sounds are dynamic, unstable, mediated, massaging the body at times while at others threatening to break its impressionable surface, to annoy the construct of one's bodily home. The sonorous envelope is thus an acoustic arena—an acoustic dimension integrating the body and the immediate world within its reverberant chamber. It is therefore highly suggestive for outlining acoustics as a question of touch, a being-in-touch. As an extended and highly volatile container, the sonorous envelope grants mediation between the force of sonic events as they touch us, as they support or upset us, defining the forms of listening and orientation one creates in navigating particular situations—the echoes we throw back, for instance, guiding us as we go. From an acoustics of intimacy to that of collective debate, from shared passions to daily rituals, from quiet mornings to noisy nights, acoustics is shaped and reshaped to facilitate or negotiate the particular social and spatial dynamics that surround. And which is fundamentally a question of repositioning one's own skin as that which is always sutured to the world.

Working around Anzieu and Ahmed, weaving questions of psychic envelopes together with queer phenomenology, affords a particular tussle with acoustics as what may enable the movement and (re)formation of togetherness—*a shared skin by which to rework loneliness and the capture of identity*. Queer acoustics is thus posed in order to *strain* a sonic phenomenology, to interrupt it with the noise of the angry or the unwanted, the disoriented and the marginalized, or the highly energetic and excited, tensing the flesh of things so as to allow for the critical articulation of resistances: to pose the work of acoustic justice.

Acoustic justice considers the practices or gestures that seek to rework the distribution of the heard, detuning or retuning the tonality of a place so as to support the movements of other bodies and voices, especially those put at risk by appearing or sounding otherwise. Lia García, a transgender artist working in Mexico City, argues that such risks are always already embedded in the sound of her transgender voice, which she refuses to modify. Rather, she utilizes the disjunctive qualities of her voice as a "noise" in order to open a social framework of affection, to destabilize what constitutes a vocal and auditory norm. Through such vocal gestures, which often require great courage, García poses her voice as a means for allowing others to "transition" with her.[11] Here, acoustic justice emerges as a work of reorientation, arguing for the mobility of marginalized voices, for voices that do not easily fall in line, that disturb and distort a dominant tonality or sonic image, and that ask of us to hear differently. For García, the question or potentiality of her voice is drawn out through her performance practice, where she often works at creating extremely moving exchanges with others through more physical and affectionate gestures, in particular within scenes undercut by histories of violence. Holding workshops within prisons in Mexico City, for example, García focuses on nurturing processes of physical and emotional sharing, between herself and a group of inmates (often with a background in gang culture), and which touch upon experiences of violence and the structures of aggression shaping prison life (and the specific orientations of masculinity). For García, her transgender voice within such situations is enabling for putting identity into question, for exposing identity as what may be reoriented in other ways, around other imaginaries, and how what it means to *be in one's skin* may be opened and shared through gestures of touch: the careful nurturing of other sensations, and other sensibilities. As she highlights, transition becomes an ontological statement that can mobilize for others a deep sense for rehabilitation, for the changing of ways, the possibility for remaking oneself and how one touches others.

Following García, and the notion of queer acoustics, struggles over the right to speak and to be heard are often about how one may stretch beyond the structures figuring oneself within scenes of violence or injury, which as Ahmed shows, is often a question of the social and material supports put into place to aid in the pursuit of certain forms of life. Questions of speaking and hearing, of voicing and being heard, are therefore equally about the vulnerabilities of the flesh, and the arrangements made to reshape how we touch and are touched, supported or hindered. In the case of the transgender voice, this may entail a struggle against the cisgender shaping of the audible. Such shaping can be understood to direct one's listening toward that which is deemed meaningful, training one to hear precisely those voices or bodies that "fall in line" with heterosexuality, the gender binary, or the straight life: that harmonize around a particular sonic image. A queer acoustics is therefore aimed at challenging such normative shaping, supporting the articulation of other images and meanings. Acoustic justice engages struggles over recognition by bringing into consideration the social and institutional structures that impact onto the politics of location and what it means to speak from one's skin. Struggles over recognition and rights are therefore positioned equally as a question of the sensual figuring of oneself in relation to others.

Musical Envelopes, Composed Noises

Questions of sound, listening, and (re)orientation find an important point of reference in the arguments posed by Tia DeNora, in particular through her ongoing research into music in everyday life. For DeNora, music offers a complex and compelling medium by which forms of bodily orientation and social relations are nurtured and managed. As she poses: "Music is, often and insidiously, part of how, for better and for worse, we are drawn into social relations and made ready, in an aesthetic and pre-cognitive way, for courses of action about which we may be otherwise (verbally) unaware."[12] Music acts as an affective, psychological, and physical material through which one finds support, aiding in arduous tasks or in easing pain; music accompanies life's rituals and routines by specifically working across interior and exterior, emotional and social life, thereby acting as a highly charged medium for attuning one's psychic-emotional metabolism and the social shape of subjectivity. Music, literally, aids in the life of the heart, giving support to the narratives by which one participates in different communities, and the urgencies and desires that drive people together. It therefore contributes greatly to defining and maintaining a sense of oneself as a body (or skin) stretched between personal and public worlds.

Importantly, DeNora suggests that music operates on a precognitive level, functioning to assuage moments of unease by setting a new pace to one's experiences; one may use music to assist in falling asleep or to bolster one in staying awake; to align with particular communities or to disrupt others; we tune and detune our relations and sense of place through musical partnerships—*this music that comes so close to me, and that connects me to others*. Within such musical formations and experiences, listening works to balance personal life with public experience over and above the cognitive and discursive articulations we often focus on. It therefore helps in modulating the meanings one gives to the more affective, sensual, and traumatic experiences of the social.

Following such observations and proposals, DeNora understands music as providing a medium or tool by which individuals undertake journeys of transformation as well. As she outlines, music "can offer an ally or template for new or emerging values or action plans. It can offer recalibrations of sensibility, and it can hold existing sensibilities in place. It can recalibrate the body in ways that call attention to the body or aspects of sensation (or, conversely, make bodily phenomena transparent)."[13] In this regard, music works to support processes of "recalibration" and, I might add, "reorientation," leading to what DeNora terms "music asylum": a type of psychic, emotional, and acoustic shelter through which to work through the ongoing challenges of being a situated subject. In moments of crisis, for instance, music is often called upon to enable a working through of difficult situations and emotions, elaborating thoughts and feelings through the particular articulations of lyrics, the sonic shape of melodies and rhythms—and by way of the embedded memories of past listening that music continuously awakens and reanimates. As DeNora ultimately poses: "Music is therefore an important medium with which to furnish (or escape from) the world."[14]

Although DeNora's notion of music asylum brings attention to music as a powerful and empowering medium for balancing oneself, I'm interested to extend music into a broader framework by focusing on sound and acoustics. This is about understanding that music, as an extremely dynamic cultural and social material, is but one type of sonic experience. And that listening, while often charged and shaped through musical experience, gives education on one's greater affective capacities to feel and share feeling, and which may be thought in this larger sense.

The movement and availability of sound, along with the complex imaginaries surrounding experiences of listening—the acoustic shape or resonance of psychic envelopes—are here emphasized as the basis for the work of recalibration and reorientation, to recognize the forceful ways in which sonic matters, and the sensitivity underpinning listening, impact onto bodily orientation, social relations, and the practices by which to furnish or escape the world—by which we grab hold and are held. As DeNora suggests, music buttresses one's physical body through any number of melodic lines, hypnotic beats, lyrical vocalizations; that is, a whole set of sounds and textures brought into a compositional form, creating a sonic vessel that ultimately carries much more than a song. Music, rather, is a type of metabolic, affective vehicle transporting sonic intensities and vibrational energies, as well as memories and meanings, that radically educates us about the dynamics of listening as the basis for a transformative process. As Paul Gilroy reminds, in a text on the black Atlantic soundscape, music first and foremost *hits* the body with its "physical power" and, consequently, one learns to hit back within one's social and political context through an array of gestures.[15] One therefore learns from music as to the intensities and potentialities of sonic experience and what it means to be situated.

Returning to Anzieu, and his notion of the sonorous envelope, experiences of sound and listening underscore the body as deeply vulnerable and susceptible to the "sound world," which includes music as one dynamic element. Music is certainly utilized, as DeNora suggests, to mediate between oneself and the experiences of location—which for Gilroy were deeply embedded in black community life in London in the 1960s and 1970s—and therefore, music dramatically inflects an individual and collective imaginary in terms of how one aligns with particular sonic materials and meanings over others; or in what ways we modulate and modify the sound worlds around us in order to work at orientation and reorientation, to forge particular alliances within struggles over "roots and routes."[16] The sonorous envelope is therefore a larger generative acoustic frame, one that integrates music as a lyrical work, as a symbolic narrative, as well as a vibrational intensity and transmitted force, which overlaps with the voices of others, of friends, the sounds in the house or on the streets, in the clubs or the schools, the flows and breaks of silence and noise—these form a greater auditory and compositional weave to which one aligns and misaligns.

This may be appreciated by turning to instances of social agitation and the expressions of particular community movements that often punctuate their struggles through sonic means. As Sirin Özgün demonstrates in her work on sounds of political actions in Istanbul, social movements and national celebrations and demonstrations find coherence and traction through "composed noises."[17] Importantly, this takes shape as an assemblage of sonic expressions and matters, from songs and music to public speeches and chanting to other cacophonous and collective noises, all of which act to support the articulation of demands or symbolic meanings beyond a particular community as well as lending coherence internally to its identities. Sounds, in this way, are mobilized, composed, and deployed to *mediate* between an internal work of community identity—a common skin—and the greater social and political systems that such actions seek to engage. The mediation of such relationships requires a fortification against *and* an appeal to others, an inward maintenance and an outward engagement, a double-work, a recalibration, to which composed noises are dramatically applied. In this sense, music asylums and sonorous envelopes are equally about the skin of a greater social body.

As Anzieu suggests, the sonorous envelope performs an acoustic mediation between oneself and the surrounding situation, registering this as a condition of immersion that necessarily unfolds the skin less as a hard line and more as an inter-sensorial membrane; where sounds touch the skin, ride upon this thin strata to impact the life of the heart, channeling all types of impressions and excitations along its surface. Sounds impart a continual registration of the world through such tactile touching, such vibrational conveyance, sensitizing one to the qualities of things around, and in return, one adjusts, rearranges, throwing sound into the world, or holding it close, defining oneself and one's place through all that may return, in the resonances and reverberations, voices and noises of others. The sonorous envelope extends the skin, unfolds it as a mediating and connective surface, affording contact by way of an acoustic performativity: to navigate around that which hinders us.

As such, the skin is but one border among many others. In defining a range of contacts, it also educates about the dynamics of borders: the rooms that open and the rooms that close upon us; the institutions that accommodate and those that shut out (that may do so according to a question of skin and its color); the social circles that invite and those that refuse. These are experiences that impact greatly onto the skin-ego, defining our capacities as individuals.

The work of acoustics, of *acousticking*, and acoustic justice I'm exploring, may allow for a questioning of borders and the negotiations that continuously pass by way of surfaces, bodies, and the flesh of things—to understand justice, and the dynamics of recognition, as always already a question of skin, of sensation and orientation, and the availability of the world to one's touch and one's recalibration. Whether individual bodies, community identities, or the territorial lines that distinguish nation-states, borders perform to delineate an inside from an outside, a scene of membership. And acoustics, as the distribution of the heard, that lends to a politics of recognition and of location, may be captured not only in terms of giving traction to particular voices but also in terms of helping to mediate between the internal, psychic life and the bodily figuring of oneself. Specifically drawing into play a question of skin, *acousticking*, and that larger arena of sound and listening it comes to mobilize, serves to enrich the work of reorientation Ahmed poses. By trafficking between the deeply affective and impressionable corporeality of the body—*these sounds that hit, this music asylum I construct*—to the social and political structures that frame us and that we may combat or capture through collective practices—*by which we hit back, this acoustic, consensual community*—acoustics brings into question the work of borders and the lines they enact.

In this sense, the sonorous envelope Anzieu poses, in conducting the movements and stirrings of sounds around oneself, and that impress themselves upon the skin and one's psychic life, performs as a container, a reverberant arena, even an auditorium: a scene by which sensing and consensus intersect. As such, it gives way to a range of acoustic challenges and affordances by which to negotiate and modulate common sense, as a unifying consensuality that may hold together a multiplicity of feelings.

Trans-Subjectivity: Libidinal Economies

What resources might we draw upon in order to support passage between interiors and exteriors, and the working through or against norms of orientation that often straighten us up? What materials or practices might be employed to support the movement of oneself, the movement of thoughts and feelings across the borders that delimit the realization and determination of individuality, and of being one with others, which may impact the regimes that work to stabilize subjectivities around particular meanings that often reinforce genealogies of violence by way of borders?

I'm interested in reflecting on acoustics and how it may lend to engaging the work of borders—acoustics, in other words, as a way of working through bordering practices and the larger acts of mediation between scenes of membership, between being hit and hitting back. Might we engage an acoustic position as the means by which to pry open borders not as hard lines or entrenched regimes of power, but rather as regions of sensation and shareability, of negotiation? As the engendering of relational conditions, might acoustics enable forms of trespass specifically in an attempt to overstep or rework the lines delineating inside from outside, the straight from the queer? From such critical and creative trespassing, what forms of (re)orientation may be created in order to engage a politics of recognition and misrecognition? That might operate, through types of vibrational and reverberant, cacophonous and stilled acts, to pull open borders and rework them as acoustic ecologies, as zones of skin-work? I'm interested to follow acoustics, from the psychoacoustic to the social acoustic, and the sonorities that are equally a question of touch, to approach borders as membranes. As such, borders emerge as facilitating contact—the borders that as bodies we carry, always already touched and trespassed by others.

I may find an expression of the border conditions I'm considering in the work of Monique Wittig, in particular in her book *The Lesbian Body*, where one encounters a body, a skin, a writing defined by an erotic tension:

> You turn m/e inside out, *I* am a glove in your hands, gently firmly inexorably holding m/y throat in your palm, *I* struggle, *I* am frantic, *I* enjoy fear, you count the veins and the arteries, you retract them to one side, you reach the vital organs, you breathe into m/y lungs through m/y mouth, *I* stifle, you hold the long tubes of the viscera, you unfold them, you uncoil them, you slide them round your neck. . . . *I* cannot speak, your teeth biting m/y cheeks your lips unscathed at the edge of m/y lips you . . .[18]

Wittig's book may assist me in getting at the skin-work I'm exploring, to support a greater understanding of borders as sensitive surfaces, territorial lines that thread bodies and identities within the larger social and political dynamics of orientation. In *The Lesbian Body*, in particular, Wittig leads us into the (un) working of borders, of language and of identity, by way of a rending of skins; through the erotic intensities of a radical consensuality, skin is brought into a performative articulation, as that which can be pulled, stretched, but which also resists, tenses. *The Lesbian Body* narrates a forceful coming together, showing the dynamics of the skin as a libidinal economy, as what strains and expresses the desires of a situated self.

Her writing steers toward a destabilization of linguistic and bodily form, a destabilization of the terms defining (patriarchal) signification, to open up to experiences of fragmentation, disturbance, as well as celebration and erotic togetherness—an erotic drive that is both pleasure and pain, adoration and revolt.

> M/y saliva spreads over your breasts, long fragments of flesh separate from the muscles falling over your neck staining your white throat, carefully *I* take them between m/y teeth, *I* chew them voraciously, then *I* look at you and *I* am overwhelmed with great pity to see you so mutilated deprived of both your arms your bust bloodied. [. . .] *I* begin to eat you again as fast as *I* can m/y so adored one *I* lick the last scraps on your belly, *I* get rid of the traces of blood, *I* absorb you m/y very precious one, *I* retain you within me.[19]

While Wittig's book is an expression of her larger project of combating what she terms "the heterosexual contract," and stakes out a militant queer position aimed at abolishing the category of "woman," *The Lesbian Body* equally forges a unique *literature of the flesh*, suggesting an erotic theory of subjectivity and of power. In this sense, I'm following Wittig as a guide toward thinking about queer acoustics, and the work of reorientation Sara Ahmed poses, as what's at stake in skin-work, or the working through of one's bodily figuring, and at the same time, to draw from Wittig an opportunity for identifying the body as a site for addressing the more general question of borders: how bodies—situated, tensed subjects—negotiate the libidinal and social economies. By way of Wittig, the flesh of things becomes a battleground, a micropolitical economy tensed by categories of gender and sexuality. *The Lesbian Body*, as Wittig suggests, is both an act of lesbian celebration and an articulation of a more "universal" project: to break the heterosexual contract in order to fight for a broader emancipation of identity itself.[20]

Skin, as the body's sensuous lining, and as what is pulled into identity, articulates a certain vulnerability, a sense of being exposed to the intrusiveness of others which, as Wittig shows, may act as the basis for a type of erotic force—a condition that enables not only exposure, or penetration but also a process of negotiating the socialization that figures one as a body, as a gendered subject. The articulations of desire and love, and the erotic charge of her literature of the flesh, capture the ways in which the body is always already in touch, touched and touching, as immersed within the physical, social, and inter-sensorial envelopes that surround, and that give passage to a possible consensuality. The dynamics of tension, of expansion and contraction, then may be emphasized as what defines the border as such, presupposing a sense of trespass, breakage, and weakness to its seeming impenetrability. A border, in other words, is always already tensed by what it attempts to hold back. The lines that orient us—lines of language, of gender—thus may become the very means for working at reorientation.

As Anzieu suggests through his notion of the skin-ego, borders are necessary precisely to *conduct* the meeting of oneself and others—to delineate inside from outside, to contain so as to protect oneself—but also to orchestrate contact and relations with the world and with others. For Anzieu, one is but the life of a border. "The skin-ego is the original parchment that acts as a palimpsest, preserving the crossed-out, scratched-through, over-written drafts of an 'original' pre-verbal writing made of traces on the skin."[21]

We travel with our skin; as a *palimpsest* of experiences, it is a carrier of memory and of history, of past inscriptions that are read by others; skin shows oneself as having lived, as having been in the world: that one has been touched, scarred in fact, by the sharing one gives or draws back from, by the dynamics of consensuality, and what one consents to or not. In this sense, the skin performs as the threshold upon which oneself *takes place*: I am written and overwritten as I move into the world, as the life I come to live unfolds, aged by the relations that define me—that mark me and through which I am myself. Through all the impressions and intensities, contacts and conversations, one gains definition—is delineated, inscribed. The skin matures as we go, and as such, it is by way of the skin and its permeability, its tension or slackness, that one's future self is worked out, as a *writing of the flesh*.

I'm interested in understanding the ways in which borders become the very site or surface by which constructs (and categories) of selfhood, and by extension, nationhood, are lived and worked through, precisely by way of contact with others—by way of the tensions always pulling at the skin. We may find indication of this in *The Lesbian Body*. Wittig's book is an articulation of the skin—not only as a corporeal surface but also as a border perennially breached, poked, captured, and overworked. The text stages an erotic destruction of subjectivity, a wearing down, a loss that additionally figures an attempt at retrieval, recuperation: resuscitation—to queer the economy of desire and the languages that shape us. Hers is an adventure story, one of skin as an erotic matter that surrenders to as well as resists penetration; it is flexed, malleable, and yet it pushes back—it tries to keep its shape under the force of the other. And yet, as Wittig shows, this is precisely where pleasure is to be found: in the erotic pull of togetherness.

As Anzieu suggests, the "body's wrapping," that is the skin, provides a primary means of self-preservation, a sense of security figured by way of the rather vulnerable tissue; yet, this containment, this wrapping, is what enables the experience of pleasure, that is, the experience of touch, of being touched, of caress: skin is the very source of pleasure, of sensation and excitation that unfolds and opens the body—a libidinal economy by which one is written and through which one works at rewriting, or directing its flows otherwise: to find that which truly brings pleasure.

Wittig's lesbian body is staged as a libidinal force, of breakage and reconfiguration, of action and escape ("Lesbians are runaways, fugitive slaves"[22]—they exit the gendered order). Her book organizes a state of desire—a radical consensuality—to delve into an experience of rapture; moreover, desire is made forceful within a larger act of reorientation, articulating an erotic queerness, "an irresistible force . . . a spinning wildly."[23] A literature that reworks language as an extension of the skin and that seeks to upset the logic of the phallus defining the symbolic patriarchal order, turning us instead toward the world of feminine flesh, an excessive flesh capturing other (textual) orientations and forms of togetherness: violent, unforgiving, possibly the first togetherness truly had or made, *the beginning of ecstasy*. What I may additionally call *togethering*, where fragmentation and joy converge to figure another sense of the body, another bodily (literary) schema "by which to reject the normal and the natural" in favor of "a thousand possible combinations."[24]

Wittig's writing comes to occupy the border—as the site of bodily orientation, inscription, of language as it territorializes, as it differentiates—precisely as a worlding endeavor; a world emerges within her text, a mythical, queer Amazonian world that twists the borders of the body into scenes of love and rebellion—where the lesbian body is posed as an erotic weapon aimed at the abolition of the construct of "woman," and that forces into the field of meaning a body of thresholds, a state of abrasions, a violent desire. "[*The Lesbian Body*] understands, in other words, that subjectivity develops from ecstasy."[25]

I follow Wittig's work in order to consider the ways in which the body is articulated as a border: as a skin, a skin-ego shaped and tensed by the libidinal economies that orient desire in particular ways and, as such, underscore the skin precisely as the site or surface by which reorientation, as a work, a struggle, an erotic possibility, is manifested: where consensuality may work at an erotic lesbian commons. Returning to Ahmed, and the notion of queer acoustics, touching and being touched are more than experiences of shared sensation; rather, they emerge as intense possibilities for realigning one's place among others, engendering another mode of subjectivity, a rearranging of the flesh of things.

Following Wittig, and the sonorous envelope mapped by Anzieu, the state of erotic intensity suggested here may lead to articulating the body as *trans-subjective*: a hyper-sensuous schema that confers upon subjectivity an erotic strength. Such a schema may allow for considering an acoustics of intimacy and affection, as that which supports an ecstatic act, in which the relational affordances found by way of the sonorous envelope come to animate the skin as a highly sensitive construct of touch: an affective radicality. Trans-subjectivity is conceptualized then as part of the tactile sonorities and meanings that enable one to *furnish or escape* the world along with and through others—by way of a bordering act, putting all of one's attention along the threshold of oneself, within the dynamics of contact. Acoustic relationality is therefore not only about extending outward, aligning with distant territories or the world of matter. It is equally about the skin, about economies of desire, sensation, and affection, capturing an erotic power: the vibrational sensations passing between bodies, the whispered words against the ear, the threaded vitality of muscles flexing, loosening, easing between oneself and another, finding expression in the movement of shared rhythms—this crafting of ecstasy. An acoustics of intimacy and affection may act in nurturing the erotic strength of consensuality, shaping narratives of the heart and the careful touching one may give. A skin-work that figures agency precisely through shared affection—a trespassing, a *trans-passing* done together.

Wittig seems to stage the body in the midst of the queer work of reorientation, pointing at the ways in which one searches for support and assistance from others; for reorientation is an arduous task, a painful ongoing act, a turning away, often, from family and friends, from what one has known thus far. A tearing apart of the normative routines and categories which one is often called to uphold, unwittingly, without question, through one's behavior and speech, through one's orientation and one's skin: how one engages or retreats from scenes of sociality. To break away, to *trans-pass*, to work toward another orientation, is a lonely project, and yet, support is found precisely through others, through other styles and stances found and adopted—realigning and synchronizing one's movements and desires to others. One works at the movements of the body, its gait, its positionality, and the expressions that attempt to redefine what may count: to find pleasure, finally. One looks for ways to reposition oneself in order to touch and be touched by that which one loves or longs for—to stretch the skin toward other fabrications and formations, other meeting points. Reorientation is thus performed upon the edges of oneself. It is a fight waged by *trans-figuring* the ways in which one is situated within particular economies and communities, along certain tracks and borders.

In his work, *In a Queer Time and Place*, Jack Halberstam charts out the importance of putting into question the heteronormative conceptualizations of time and space, which often shape our capacities to live otherwise.[26] From experiences of childhood and adolescence through to adulthood and experiences of aging, life is contoured and channeled according to a dominant, heteronormative shaping of time lived and space inhabited. Halberstam works at reorienting life around other conceptions of time and space, in which queer temporalities and spatialities may enable other ways of being in the world. As he outlines, "In Western cultures, we chart the emergence of the adult from the dangerous and unruly period of adolescence as a desired process of maturation; and we create longevity as the most desirable future. We applaud the pursuit of a long life (under any circumstances) and pathologize modes of living that show little or no concern for longevity."[27] In contrast, Halberstam maps out queer temporalities and spatialities that work around other conceptualizations, for instance through "stretched-out adolescence," as well as queer constructs of "family," which lead to other life narratives.

Queer temporalities and spatialities are mobilized in order to support the articulation of queer life, as what fundamentally works against or around heteronormative conceptualizations of the stages of life and the forms through which they are expressed. Additionally, queer temporalities and spatialities give support to the work of reorientation posed by Ahmed: as a work one does in seeking out other forms of alignment, as a retuning of the compositional orders in which one is placed. As Ahmed argues, phenomenal matters are deeply shaped by existing normative constructs that precede one's arrival and that impact how one may come to occupy and flourish through such constructs. Within the scene of touching, of being touched by the matters surrounding, Ahmed, and Halberstam, seek to capture a potentiality, the nurturing of the capacity to live otherwise, to reconstruct time and space so as to honor the particular joys and imagination, experiences and desires of queer life.

The desire for other orientations, and the transfiguration of bodily positioning as the opening toward other temporalities and spatialities, is further considered by the writer and theorist Nina Dragičević. Dragičević researches the culture of queer communities, particularly in the city of Ljubljana where she lives, giving attention to the medium of sound and the acoustic dynamics that shape the social environments of gay bars. As she suggests, sound and music, listening, and an overall acoustic dynamic contribute greatly to supporting queer togetherness, particularly when speaking out loud may put one in danger. Rather, the articulation of desire turns upon a sonic axis, a queer acoustics, finding facilitation through the playback of particular music: songs, in fact, assist in narrating these otherwise unspoken communications, where potential partners may stand in or identify with singers, or those being sung to. Songs, in this sense, provide an acoustic affordance enabling the expression of desire, and importantly, for the construction and maintenance of a culture of queer life.[28]

A queer acoustics may be found as part of the making of worlds, facilitating the generation of shared temporalities and spatialities, a trans-passing to aid in the reconfiguration of particular environments, identities, categories, and communities, and by extension, oneself. To resound, to retune, to figure a passionate dissonance against the drift toward certain consonances, queer acoustics may assist in negotiating the underlying heteronormative structuring often defining spaces, from homes to bars, schools to workplaces. Finding solidarity with others through the intensification of the sharing of skins acts upon the dynamics of a sonorous envelope and its libidinal economies, allowing for specific reverberations and transmissions that necessarily impart their erotic meanings onto those relations.

Detention: Rhythms of Power

I've been pursuing questions of acoustics and acoustic justice by way of the skin, and the ways in which the joys and struggles of togetherness often necessitate the difficult work of reorientation: of carving out or fighting for particular temporalities and spatialities. Within such processes, such tensed negotiations, acoustics is positioned as a framework for appreciating how sound and listening, attunements and alignments, afford acts of mediation, rebellion, *consensuality*, acoustics as a deeply affective means for furnishing or escaping the world—for recomposing the order of things. From my perspective, acoustics draws out "the border" as a contested line, a fleshy resonant line, giving assistance in contending with the delineation of identities and the powers that situate them. From the voice that moves one into scenes and situations to the gestures that attune or interrupt, to inaugurate the synchronization by which communities are often shaped, a *betweenness* emerges, that is materialized. Betweenness, though, is a difficult position; it both supports and fragments at the same time, operating to connect while also pulling apart—this that binds, tethers *and* stretches, disintegrates. Betweenness is thus not a soft ideal of commingling, but rather a state of force. It is not a "third term" resolving conflicts, but rather a continuum, an erotic strength. It is a power to be carried into the work of reorientation, in the trespasses and trans-passes undertaken.

Wittig shows ecstasy as such a state, a vulnerable and dynamic state—a *being undone* by both desire and death—"*I* open m/y mouth, *I* admit your lips your tongue your palate, *I* prepare to die by your side";[29] a state constituted by both pain and pleasure, a pain/pleasure held in tension, and that plays at the skin, pulling too hard or letting go, where intimacy is dynamic, trans-subjective: a bodily figuring flushed with desire, drive, affection, fragmentation, and thus requiring negotiation, navigation, responsiveness as one is pulled into another. How might one work through such processes, developing a skin-work to modulate the dynamics of touch? To figure an acoustics—as an economy of attention, of libidinal drive—to work through the skin, as a site of tension, of betweenness; to configure a queer acoustics as what may contend with the coordinates of time and space, the timings and spacings, rhythms and vibrations of the sonorous envelope, to give way to a world of pleasure, escape?

Accordingly, acoustics and acoustic practices may be elaborated as a base for the working through of borders, as what may give support and protection while affording contact with others. The highly malleable and energetic matter of sound—as vibrational and tactile, the whispered word or the noise you make, in the sympathies and interruptions by which bodies are *enlivened, agitated*—lends to how one may position one's skin precisely as what touches the world and others, what figures by way of a continual propagation, dissemination, a betweenness, this state of being undone and redone. Betweenness thus requires attention, generosity, withdrawal, a skin-work that pushes and pulls at the lines of separation, of difference. Betweenness, in this regard, is a position one takes in being oneself, a being oneself by way of others, by way of the ecstasies of being-in-touch.

In following these notions of skin and borders, it becomes important, if not essential, to consider acoustics not only as what may enable scenes of joyous work, an acoustics of intimacy and ecstasy affording shared worlds, but equally how it comes to perform as part of systems of oppression or denial. The skin is additionally captured within the border regimes that utilize profiling, locating race, gender, and ethnicity as key features in governing the movement of people. The skin-ego mapped by Anzieu, performed and staged by Wittig, is forcefully positioned and captured within a larger context of global borders. Here, the borders of the skin helping to shape and protect subjectivity, to conduct the joys and ecstasies of togetherness, are brought into the project of national identity and the securitization of the nation-state.

As Kim Rygiel argues in her work on migration, contemporary border regimes enact forms of global governing that exceed the physical borders of nation-states; rather, borders are now firmly placed *onto* the body. Through an intensification of border security, new technical and ideological apparatuses, such as biometric technologies, data mining, electronic tagging, and identity profiling, now register the body as a carrier of risk: "Detention operates not just through spatial containment but also through governing mobile bodies as they move through and across space. In the latter case, governing mobility occurs not through the spatial containment of mobile bodies but, rather, through tracking mobile bodies as they move across space."[30] For Rygiel, this is indicative of a greater transformation of citizenship in general, and understandings of its relation to the nation-state. Citizenship, under the continued and complex push of globalization, now functions as a form of government, where nation-states collaborate across borders to manage the physical mobility of people, especially those from the Global South.

> From a citizenship-as-government perspective, the transformation of the sovereign state in response to globalization processes does not mean the erosion of modern citizenship as an institution and governing regime. . . . Rather, from this critical perspective, globalization may be seen as transforming citizenship in ways that disrupt its close association with territorial borders . . . [to appear] in more exclusive ways. Strengthening it to make it a more powerful regime for governing individuals, populations, and their movement between and across borders.[31]

As part of this new citizenship-as-government, nation-state borders are no longer policed at their physical sites only; rather, borders follow bodies as they move, distinguishing from citizen and non-citizen, from those with access to rights and those without. Such a process greatly disturbs those seeking asylum and refuge. "The interruption, through detention, of the ability of certain individuals and groups to acquire the spatial association and presence within a political community from which to then make claims, disrupts their ability to be and become citizens, with the right to have other social, economic, political, and cultural rights."[32]

Within the new apparatuses of global control, biopolitical power is pressed onto the skin-ego Anzieu poses, giving way to the emergence of particular "abject subjectivities" and their contemporary policing. Specifically, new forms of detention practices work to govern by restricting the movement of people, especially those escaping experiences of oppression and violence; refugees, asylum seekers, and undocumented migrants are subject to new modes of detention practices, and are often placed "on hold" as they navigate through a range of national and transnational procedures and protocols.

Detention practices are founded on spatial, temporal, and technological apparatuses that explicitly capture migrants' and refugees' abilities to set pace to their own rhythms and movements. Numerous restrictions on accessing education, work, health care, not to mention greater social and cultural frameworks and communities, perform to control migrants' spatial and temporal experiences. As such, contemporary security regimes work at aggravating the plight of migrants and refugees through the conditions and experiences not only of detention, but of waiting: a going nowhere that saturates and frustrates one's emotional and psychological life. Processes of detention emerge as powerful devices within biopolitical governance, operating to deter the flow of refugees and migrants by installing a mechanics of waiting—a prolonged limbo that cuts against any sense of hope. As Synnøve Bendixsen and Thomas Hylland Eriksen outline in their work on migrant struggles, waiting is figured within a larger operation of detention, and can be understood as thwarting feelings of hope—that sense for moving away and moving toward, from a past of often violent and traumatic experience toward a future of a better life, of security and safety. While "active waiting keeps open what is anticipated, and therefore entails hope understood as generative of action,"[33] waiting as experienced under the directive of detention regimes evacuates one's sense of activity, of being in the midst of an articulated temporality. Through such inactivity, one experiences a deep sense of having "lost control over one's own time."[34]

Waiting by way of detention is a type of cut, a suspension of the rhythm of activity and a subsequent prolongation of one's feelings of loss; a draining of that energy that is hope. "As long as there are things to plan for, things that one believes or hopes will affect one's circumstances, activities seem filled with purpose and direction."[35] In contrast, detention often deepens that profound psychological sense of deadness, where time is nothing but a weight, where the sense of temporal passage and direction is suspended, lost. A going nowhere, a hopeless passing of time that strips one from capturing a sense of purposefulness.

I want to reflect further upon questions of temporality, and the force of waiting that often define experiences of detention, especially that of indefinite detention. To do so, I turn to an archive of audio recordings resulting from the Manus Recording Project Collective launched in 2018 in Australia. Connected to a larger research and curatorial initiative on sound and the politics of listening led by scholar James Parker and curator Joel Stern from the cultural organization Liquid Architecture based in Melbourne, the Manus Recording Project Collective initiated a process of collecting and transmitting daily audio recordings produced by a group of participating asylum seekers and refugees held at the processing center on Manus Island, Papua New Guinea, as part of Australia's offshore detention and immigration processing infrastructure. Developed in collaboration with journalists and radio makers André Dao, Michael Green, and Jon Tjhia, the resulting archive of over fourteen hours of recordings provides an important view onto experiences of indefinite detention.[36] Presented as part of the exhibition, *Eavesdropping*, at the University of Melbourne's Ian Potter Museum of Art, the audio recordings (presented under the title "how are you today") were uploaded daily to the Museum over the course of three months and made available to visitors through an installation of audio speakers, forming an unedited reportage by detainees.[37]

The Manus Recording project sought to intervene within the dire conditions experienced by detainees, allowing them a mediated form of contact and communication with the outside world. Many of the detainees had been in detention for over five years, having been picked up on boats traveling from Indonesia and with no prospect of ever settling in Australia.[38] As such, the activity of recording enabled detainees opportunities to tell their stories, and to inform a larger public about their experiences and the deeply problematic approach taken by the Australian government as to the treatment of refugees and asylum seekers.[39] The project additionally articulates a sense for interrupting the detention practices and policies on the part of concerned citizens—to protest by way of the project, creating a space or mechanism by which to reach those being held on the Island, to capture a sense of solidarity and collective resistance. The recordings thus act as a form of reporting as well as lending a sense of agency, however indeterminate, to those held inside and those who disagree with national policy.

To speak and to be heard, by way of recordings sent across closed borders, is a powerful act, one that mobilizes the urgency and potency at the heart of hearing: the sense of support and activism garnered by way of being listened to, and how listening may in fact articulate a position of protest: to hear what otherwise can't be heard. Importantly, the recordings also reveal a more nuanced view onto the experience of detainees. "Reporting" here is not necessarily an articulation of well-formulated views; rather, the recordings *expose* the daily life of the detainees, capturing an extremely textured view onto their reality. Forms of transmission, in this way, are deeply effective, granting one's experience of displacement and captivity a potential form of reprieve, a sense of there existing a somewhere else, beyond the emptiness of indefinite detention. Transmission may wield an intervention onto the scene of closed borders, acting to manifest forms of contact and support.

In addition to the sense of possibility and protest opened up through the recording and transmitting process, I want to reflect upon the ways in which the recordings capture and make tangible the dynamics of waiting that pervade the life of the detainees. A deep feeling of suspension and emptiness texture the recordings, and often the detainees highlight this in what they say. For instance, Samad on July 25, 2018, squarely states that: "we have nothing to do here." In response, he immediately starts to talk about music, and how music is something that helps him during his days of waiting. As he tellingly offers, "music is my close friend." Music is highlighted as an emotional carrier, even a channel toward feeling. He continues, "music is the only thing which makes me feel better, sometimes it makes me cry, sometimes it makes me sad, sometimes it makes me happy." Following this, he commences to play a recording of a song that he likes, especially what he calls "slow music." In playing the song, recording this act, and sharing it with us—we who are only ever imagined, as a projected sense of others who are listening—we're brought into a temporal dimension together: taking the time to listen to this slow music with Samad, one begins to sense the duration that surrounds him as a detainee. Time is brought forward in this instant of shared listening, an impossible listening performed by way of recorded sound, one that enables a form of *being-with*. This is clearly a powerful quality of recorded sound in general, the ways in which it may intervene within a given time and place to overlay a sense of the original moment onto the present. I begin to feel the recording not only as sonic matter, or as representational, but also as an experience that was once lived by another, Samad, which upon playback is brought into the present, almost resuscitated: my listening, now, as the resuscitation I give, or am obliged to give. Recorded sound may act as a form of emotional archive, a set of held frequencies that upon playback take up residence, suddenly, within the time and place of the moment, shivering the air with its touch. This is precisely part of the power of music in everyday life that Tia DeNora maps: music asylums are most often found by way of recorded music that we carry with us. As Samad suggests, music is a friend precisely because it remains close. Such availability, amplifications, and transmissions wield a great power to carry meanings, and to animate or transform, with a complex emotional and social energy, the status of the present.

Throughout the Manus Island recordings, explicit references to music are continually made, either through listening to recorded songs, or singing and learning songs on the guitar; music becomes an extremely potent partner and theme throughout the project. We must appreciate how music emerges, is performed and listened to, affording an emotional and psychological channel or "scaffolding" for easing the detainees' sense of temporal and spatial displacement. Music, as DeNora notes, becomes useful within scenes of crisis, impacting upon one's psychological state by offering an emotional partnership, a form of support.[40] In following Samad, as well as others, such as Aziz or Farhad, music clearly works by enabling a sense for being able to ease one's suffering and loneliness. Often music is played in order to "return home," recapturing a sense for what was left behind, for the material and emotional bonds of life before detention. Importantly, music may be heard to enable the detainees the capacity to define their own time. By choosing what to listen to and when, the detainees personalize their environments, their spatial and temporal situations, reclaiming a sense for crafting the way they inhabit the present and what they may carry with them—memories of home, of life lived prior to their journeys, and that are given renewed traction through the hearing of music close to their heart.

The time of waiting, of temporal suspension, is reshaped into another form of duration through these musical experiences—these experiences that are explicitly crafted to reorder the psychic, sonorous envelope, to figure by way of sound another time and space: another home. Music gives access to time in an extremely palpable and material way. Time, in short, becomes *active*, tangible, with the detainees filling the minutes with an energetic presence, an emotional and physical sensation that makes one feel that time is not wasted. Rather, time becomes productive, or rather, it becomes one's own—it becomes lived time. Here, within the brutal scene of biopolitical capture and governance, I'm led to hear music as that which fuels a sense for controlling one's own body, returning one's skin to oneself through sensing the vibrations of a song, this song that makes Samad sad—*that makes him feel himself*. Under the push of a song, Samad intervenes within the timing and spacing of detention, capturing these as materials, a chamber that may start to resound with a generative acoustic, affording the movements of an altogether different sensation.

The experience of *feeling with music* leads one back to the body, to the story of oneself; music enables a telling of oneself—a fleshy, moving, and feeling self, a person with history. Returning to Nina Dragičević, songs often enable the articulation of desires that cannot be otherwise spoken aloud. They pick one up, transmit and transfigure; songs also heal, acting as a musical moment that one may always return to—that one may sing along with, as the detainee Farhad does while cooking. One may rely upon music, as that which grants physical and emotional support, enabling one to furnish or escape the world—to mirror back one to oneself, confirming where one has been and importantly, where one may still go. The affective and kinesthetic qualities of music, whether in the playing of instruments or in the moment of dancing and singing—or with Farhad, who at moments cooks in time to the music he listens to, and even, whistles its melodies—these qualities are brought forward within the scene of biopolitical capture to ease one's time of waiting. In this sense, waiting is recomposed into a state of vibration, a "practice of vibration" that, as Nina Sun Eidsheim poses, produces "inter-material" articulations: a resonance, a *singing along with*, that passes across the body and the material of the world to engender a power, a shared skin, which may exceed the capture of oneself, at least for the duration of a song.[41]

Other Temporalities, Other Synchronizations

In his lectures on the topic of living together, Roland Barthes considers what he terms "small social formations": the assembly not of mass demonstrations or urban strikes, nor of societies or states, but rather of groupings of individuals. Taken from cultures of monastic life and their communal orientation, he poses "idiorrhythmy" as a way to probe the relation between individuality and community, self and the social. Idiorrhythmy is mobilized to capture how personal rhythms inflect or brush against a greater order of living together, tensing this order with its syncopations, its elations and depressions, its abilities and inabilities. In short, Barthes draws upon idiorrhythmy to imagine models for being alone *and* being a part of life with others at the same time. In her preface to the related publication, Kate Briggs emphasizes how idiorrhythmy "names any attempt to reconcile collective life with individual life, the independence of the subject with the sociability of the group."[42]

In thinking through how to live together, Barthes maps a range of spatial formations, from small enclosures and huts to prisons and monasteries, reflecting in a highly diagrammatic fashion on the forms that shared life might take. In this regard, Barthes captures the ways in which time and space are interlinked, brought into an expressive articulation. Rhythm thus names a form of composed time, one which immediately draws out a question of spatial arrangements, of spacing, and of architecture as contributing to the articulation of movement. The production of space, as Henri Lefebvre reminds, is ultimately a question of social life and the rhythms defining it.[43]

Within this framework of rhythm, of time and space shaped into forms of life, worked over as material, idiorrhythmy is positioned between the life of the body and the form of a social order, a rhythmic pattern whose dynamics register this relation as a performative temporality, a spacing tensed by negotiation. As such, idiorrhythmy is what grants room and flexibility to the regimentation of social ordering and the grounding of identity within a field of meaning. Idiorrhythmy, rather, cuts against the project of identification as a dominant timing or synchronization, giving flexibility to the movements of subjectivity by figuring improvisations, lapses, intervals, "queer temporalities and spatialities"—in short, an interplay that gives way to drift and the pleasures of movement itself. As Barthes suggests: "Before anything else, the first thing that power imposes is a rhythm (to everything: a rhythm of life, of time, of thought, of speech). The demand for idiorrhythmy is always made in opposition to power."[44]

In reflecting upon idiorrhythmy, as that possibility Barthes poses for crafting individual movement, for figuring a gap within social ordering, waiting seems to pose a problem. As Barthes suggests, idiorrhythmy indicates "a flexible, free, mobile rhythm" that results in "a transitory, fleeting form, but a form nonetheless."[45] As such, idiorrhythmy is an easing up of regimentation, a giving way to flexibility, even improvisation; it never leaves the order of social life completely, rather, it signals individuation, as a duration of moving otherwise—a "form nonetheless." Within Barthes' model though, indefinite detention, as a form of inactive waiting, seems to suggest another view. Here, waiting seems to explicitly work at nullifying rhythm—there is no definition here, no structure against which one may move; there is no beginning or end ordering time and around which individuation may figure itself, may resist or accommodate, may appropriate or refuse. Indefinite detention edges toward formlessness, a nothing, a type of oblivion: time suspended and rhythm denied.

As Henri Lefebvre poses in *Rhythmanalysis*, rhythm is the result of the interactions between "time, place, and energy";[46] from out of such interactions and intersections, rhythms emerge, giving indication as to the dynamics at play between time and place, and the vitalities that move and thus register history, memory, sociality—that is, the bristling of life lived, relations made, time carried. Waiting, under the directives of indefinite detention, seems to upset Lefebvre's equation, for where is time and place within the detention center on Manus Island? And where does energy reside, emerge, or figure within the rooms in which nothing happens, where inactivity is prolonged as a state of being? Time, place, and energy align into another type of configuration: within an indefinite duration, rhythm is blocked or barely articulated. As Lefebvre suggestively states: "Rhythms always need a reference."[47] Rhythm relates itself to a given scene, a situation to which it moves, is moved into expression; it pushes against, emerges from within, this rhythm. Within the scene on Manus Island, references are greatly reduced if not violently evacuated.[48]

Waiting, and the current practices of detention performed between governments of the Global North, comes to define a spatial and temporal injustice (among others), imposing an indefinite absence onto the life of detainees. Here, power establishes a rhythm not of regimentation, but of suspension. Such practices require deeper understanding as to the shifting dynamics around what Kim Rygiel terms "citizenship-as-government," where border regimes no longer reside at the territorial lines of nation-states, but rather invade the body through a biopolitical governmentality: a monitoring of the racialized, gendered, and ethnic body whose movements may be cast as threatening to the sovereignty of a larger transnational order of the North. Indefinite detention, as waiting prolonged, as time suspended along with the rhythms necessary for punctuating one's life, must be understood as a form of violence.

Following Barthes' speculations, his diagrams of rhythm as a spatial question, as a frame of relations, I want to return to the Manus Recording project in order to reflect upon the experience of listening they afford, and what the recordings offer in terms of exposing the life of detainees and their experiences of waiting. The Manus Recording project renders for listeners the experience of time inside: as detained time. Registered in the daily reporting, which performs a type of counting (in this case, from July 24, 2018 to October 28, 2018), the days come to emerge: a counting that works at punctuating the time of indefinite detention on the Island. The structure of daily reporting begins to form a diary not only in terms of giving expression to particular narratives but also by making palpable the experience of time, and more: the counting of time at the core of detention. In this case, counting wages a fight against the indefinite; against that sense of endless time, detained time, counting begins to make time matter by indexing it, registering it within a durational frame. It begins to provide a reference: a referencing that turns into a pattern, a record, a story. The experience of time inside is further articulated through the recordings themselves, produced as ten-minute blocks in which *something happens*: someone plays guitar, another listens to music; someone cooks, another speaks, or we hear rain, descriptions of the Island, conversations about legal struggles; also testimonies as to the mental health of detainees, or about their hunger strikes. And always descriptions about wasting time: watching movies, playing cards, passing the days. In a sense, as listeners we are given time itself, punctuated in durations of nothing and then something, moving from the banal to the deeply troubling.

As listeners, we also *wait for* something, an insight into the untold experiences of detention, the stories and lives withdrawn. We follow the detainees, as they look for ways to survive: something that may turn this time into an opening, to reconfigure experiences of waiting toward a rhythm of hope, a sense of articulated life. The recordings thus work to fundamentally capture the push and pull taking place between orders of time: between indefinite detention and life lived outside, between waiting and figuring a sudden rhythm: a pleasure, a break, an interval by which the uncounted may begin to count.

Subsequently, from out of the act of recording (and its transmission) rhythm emerges, is brought into play to figure durations of potential ease, protest, contact, audibility; these ten-minute daily reports by which detainees may focus their time into a project: of narrating, doing something, of the possibility of sharing, escape, of touching and of being touched: of extending oneself beyond the confines. A project that turns waiting into something else: into ten minutes by which to end waiting. "The project is really beautiful for me, this is new experience. I can record a file audio about 10 minutes and share with people and they understand, what life is look like on Manus Island for me."[49]

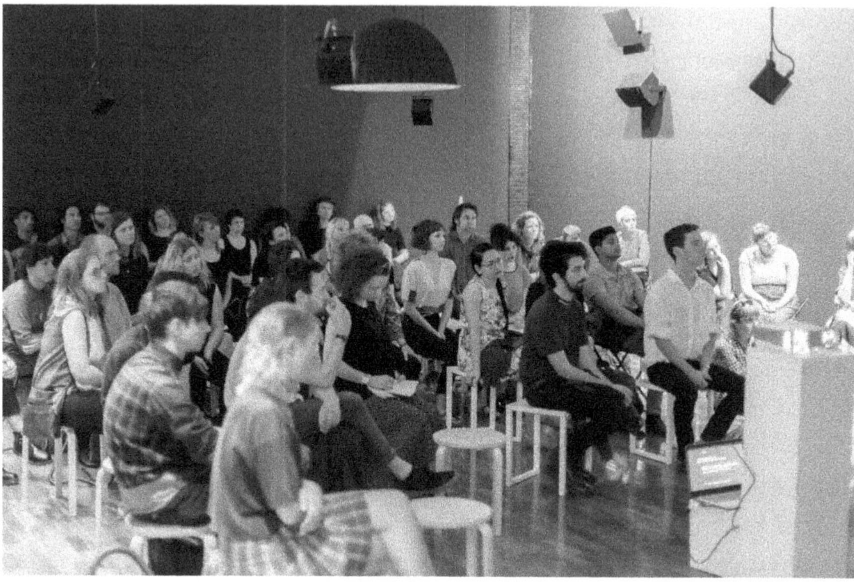

Figure 3.1 Audience in attendance for the listening event "How are you today: Listening to the Manus Recording Project Collective," Ian Potter Museum of Art, Melbourne, October 19, 2018. As part of *Eavesdropping*, an exhibition curated by James Parker and Joel Stern (Liquid Architecture). Photo: Jacqui Shelton.

The Manus Recording project offers but one type of intervention onto the scene of detention, figuring a shift in the absence of rhythm by providing a reference point: a channel of transmission, an exhibition, a sociality. In registering life inside, the recordings show us the importance of making time significant—to enacting, through the device of an audio recorder and an uploaded audio file, a gesture of acoustic justice: this small window suddenly opened, this crack out of which the configuration of time, place, and energy held by indefinite detention starts to take on substance. A sense of life or mobility may, in fact, emerge, rhythmed by listening and that sense of there being others. And in doing so, such gestures may work against the systems that frame certain lives as "ungrievable" and therefore, without value.[50]

Matrixial Narratives: Shareability

Questions of migration and the contemporary regimes of governing citizenship dramatically draw out the issue of borders. Borders are sites and scenes of contact and of negotiation, and as such, they are generative of a range of both pleasurable and painful experiences. Upon the border, a politics of encounter and recognition, passage and reunion are dramatically played out, leading to spirited exchanges among friends and strangers, as well as negotiations as to access and entry, security and safety. Entering and exiting, generosity and refusal, passage and detention exert their tremendous force onto individuals and communities as they confront the border, shifting from the personal flesh of the body to the impersonal walls of the nation-state. Am I welcome here? Am I accepted as a member here, as a person? To what degree am I recognized, and given access to social and institutional support? Brushing past or against the borders that surround, and that delineate and delimit this body as racialized and gendered, as one or the other, is to sense the heavy weight or fluid lightness of one's body.

In her research on experiences of trauma and historical memory, the artist and psychoanalyst Bracha Ettinger provides a compelling framework for thinking borders. Importantly, she works to unsettle the theoretical legacies inherited from Freud and Lacan that emphasize the formation of subjectivity as being bound to a (masculine) narrative of castration. In contrast, Ettinger attempts to rework Freud's theories of the subject by developing an alternative view based on what she terms "matrixial borderspace."[51] Whereas Freud (and by extension, Lacan) emphasizes the emergence of the subject as a cut from the maternal bond, a severing from the "matrixial" that ultimately locates the subject around the symbolic logic (and power) of the phallus (and the narrative of castration)—a persistent and dominating absence or lack, and the desire for a return; Ettinger, rather, seeks to understand the subject as being always already bound to the "feminine"—the matrixial, a womb-narrative—and as such, to a process and potential of "co-emergence" and "inter-connectivity." For Ettinger, that which is other—the difference of the mOther—does not fix the subject to a perennial agony, a constant replay of difference as that which forever separates one, as loss; rather, difference is generative for what she terms "relations-without-relating." "Here, the emergence of meaning is not related to *absence* or even to the rhythmic movement of presence and absence, but to shareability and jointedness, and to changes in and of *distance-in-proximity*."[52]

Ettinger leads us toward the formation of a subject shaped by the feminine, highlighting a matrixial specificity that steps outside the phallic logic of Lacan's *objet petit a—this (object) that I desire but cannot have, this that agonizes me*—producing instead a "border-link" or "border-object" that ties the subject to an array of others: objects and subjects, as well as losses and exchanges whose psychic charge is not held by the narrative of castration: by a cutting, a giving up. Shareability and jointedness, rather, define the matrixial-subject, allowing for another frame of relationality, another libidinal economy of desire, one that places us within a continuous contact with strangeness, a *being alongside and with* otherness. As Ettinger suggests: "A certain awareness of the *borderspace shared with an intimate stranger* and of the co-emergence in difference corresponds to a feminine dimension of subjectivity."[53]

In capturing the feminine dimension of co-emergence, a positive *being alongside* the stranger, a procreative otherness, this difference I cannot fully know yet which is close to me, that touches me, Ettinger opens a productive and enriching understanding of the subject as a co-formation, a "trans-subjectivity" where one is driven not by the agony of a continual lack, where the "womb is denied," but rather is conditioned by "*the impossibility of not-sharing*."[54] Matrixial borderspace, as she poses, is generative for linkages and co-emergences, for shareability and jointedness that ultimately works in support of compassion and hospitality: relations-without-relating as a scene of radical plurality.

Ettinger's model of subjectivity—this *impossibility of not-sharing*—may help in considering the border less as a scene of capture, of cutting, as a politics of othering, and more as one of procreation and cohabitation; whether on the borders of nation-states, or along the edges of one's own body, subjectivity emerges (is engendered) as a process of encounter, exposure, jointedness. For Ettinger then, borders are spaces in which one enacts one's humanness; it is precisely at or within the border that we become ourselves as always already bound to others, to the matrixial engendering (a womb-narrative) upon which life is based.

Returning to Kim Rygiel, and the issue of migration struggles, it becomes important that, along with the establishment of new detention practices, and citizenship-as-government, one takes stock of the subjectivities and alliances emerging from within new border controls: these globalized subjects that force into question the very nature of borders of nation-states and what constitutes citizenship today, and that often incite other narratives of globalization. For Rygiel and others, such as Engin F. Isin and Vicki Squire, citizenship is underscored less as a form of membership and more as an *act*.[55] Thus, borders—whether along the walls of the nation-state or upon the flesh of the tracked body—provide a space, a scene by which to capture citizenship as both a transnational and transitioning act. For it is upon borders that we may transition together, queering the temporalities and spatialities that work to detain "abject" (or deviant) subjects. From such a view, borders emerge as important contact zones, not as "extra-territorial" apparatuses of state sovereignty, but rather as intra-territories: from the erotic story of togethering to the building of new alliances against the directives of state governments, borders are precisely what are enacted and reformed in the giving of oneself and, as such, are never so fixed or firm. As such, they act as dynamic thresholds engendering the ongoing articulations of oneself and others as inseparable. Closing borders is therefore an extremely violent (and pathological) act.

Returning to Manus Island, and the detention practices of the Australian government, borders have become extensive, extra-territorial mechanisms that often work to deny, on many levels, the movements of those seeking to claim rights and access to a safe and supportive environment. Here, as the Manus Recording project attests, we may detect the emergence of other movements—other ways of navigating borders, as well as the traumas they may produce. Writing from Manus Island, the Kurdish journalist, poet, and detainee Behrouz Boochani underscores the incredibly resistant and hopeful position embodied by refugees as they nonviolently resisted their forced relocation by the police in 2017:

> According to its own logic, and consistent with the character it has moulded itself into, the detention regime wanted to manufacture a particular kind of refugee with a particular kind of response. However, the refugees were able to regain their identity, regain their rights, regain their dignity. In fact, what has occurred is essentially a new form of identification, which asserts that we are human beings.[56]

Boochani's letter from Manus Island, along with the audio recordings sent over the digital networks spanning the globe, shows that what we understand as citizenship is found not at the counters of passport control, but rather along the borders of one's flesh: the skin that, while being tracked and profiled, is equally pulling itself back up and into its human form—this human form that shows itself, punctuates itself, as that which is able to join together, to be shared among the *flesh of the world*. How we listen to these acts of punctuation and jointedness becomes central to the work of borders and the narratives and lives we allow them to support or deny. In this sense, acts of citizenship—whether in this moment on Manus Island when refugees reclaimed their own dignity, together, or when people traffic boats in order to pick up migrants often against the decrees of nation-states—I would suggest *empty* the border, casting it as a space, a volume, an acoustic chamber in which sovereignty manifests in the rhythms and transmissions, the sensualities and meanings passing across rather than held on either side.

As the work of Bracha Ettinger suggests, borders are less a line and more a space, a *borderspace*, a betweenness, which calls forth the possibility of inhabitation, whether along the borders of the body that, following Monique Wittig, may become a site for radical sharing, a deep interpenetrative writing, a *transfigurative* act, or that of the nation-state border, which seems to reside everywhere within today's globalized order. The borderspace, as that scene where the "impossibility of not-sharing" takes place, may start to give articulation to another type of citizenship, along with the building of new kinships—where sad songs may be shared in an effort to exit the hole of waiting, and where the yearning "for the joys of childhood, for mystical movement, for freestyle rhythms, for liberation through dance,"[57] which Boochani experiences, may find expression.

Communities in Movement, Communities in Transition

How might we think further about acoustic justice in relation to questions of subjectivity and community, borders and bordering? In what way does the performativity of subjectivity, of personhood, include or incite an acoustic question, an engaged relation to hearing, feeling, as well as the work of reorientation? Might we find by way of acoustic understanding a model by which to foster particular temporalities and spatialities so as to assist in negotiating forms of governmentality—to cultivate what Judith Butler calls an "egalitarian imaginary"?[58] While borders and boundaries are necessary in the preservation of the self, in struggles over (marginalized) identity and the politics of recognition, I'm keen to approach such struggles less through acts of fortification, less through a project of identification, and more through understanding permeability as a type of erotic strength, a passionate strength, for staying with the complex dynamics of coexistence. The work of borders, as I'm suggesting, is predicated not on keeping out, nor on keeping back the tendency of movement; not on practices of policing, but rather on communicating, in figuring or facilitating conduits; a labor fixed not on posing contact as always already a threat, but rather on spiriting methods of negotiation. In this regard, understandings of agency are not strictly defined by way of self-determination, but rather find meaningful expression through strategies of generosity, of care, even of love, which underscore power as what emerges and is held by its distribution and sharing.

As I've tried to highlight, this entails appreciating subjectivity as always already *in touch*, as trans-subjective, and which entails a negotiation with existing alignments and synchronizations; subjectivity as oriented by an externality that is always closer than imagined, that supports and undoes at the same moment: that places oneself *on the skin*, which as Anzieu reminds, is always greater than the physical limits of the body. Rather, skin is a question of an extended touching: abrasions and caresses, pathways and closures, bonds and breakages that define a relational understanding. Skin, in fact, as what shows us *as* a vitality, as being who we are, putting us in tension with others, and therefore, within the politics and poetics of contact, the togethering we may work on. The border, in this regard, is precisely where my vulnerability and my strength are exposed, defined, and where I am most in need of support, and through which I may also lend support. Betweenness is therefore the risk one takes, in caring, in being with others, in loving, and as such, must be taken.

This is not to overlook the degree to which some skins are more vulnerable than others, and that taking risks is never an evenly distributed capacity that all bodies may share—taking risks is seriously more risky for some more than others. It is not my intention to overlook or diminish the uneven politics and powers defining the skin—for the skin, as Adrienne Rich reminds, is never a neutral affair. It is, rather, because of this that I'm seeking to *address* the skin precisely as what is stretched, and in being stretched requires greater care, a *critical dressing* for those wounds that one carries, and that one may also inflict, knowingly or unknowingly; and that may also lend to how to *redress* the coercive figuring of people at risk.

In this way, I may ask, what skins resist my own touch? Whose bodies am I addressing in this discursive act, and do they, in fact, need my discourse? What are the limits of my address, and what limits should I bear in mind when reaching toward such precarious and violent scenes as found, for example, on Manus Island? Who am I to write about Samad, or Boochani, for example, as they live a life I can only ever imagine from a distance? But which I may start to sense, along the borders of my own skin as it is pulled by way of an audio archive in their direction.

In this regard, I take my cues from others: from the examples, the words, the reverberations of other voices as they touch me. And to which I feel drawn to respond, to move while remaining responsive to all that I receive as well as take. In short, I understand the limits of my discourse, the limits of my own body, as that which connects rather than separates, that figures me accountable, responsible, shareable, and therefore vulnerable: a vulnerability that acts as the basis for a critical strength, for a being-in-touch, which is always a tension—with other skins that pull away, that burden or excite with their pulling, and which I follow into other discourses, other worlds, other strange and unfamiliar experiences: especially toward that which I may never truly know, but which I may certainly learn from, and possibly even come to embrace. A commonality that is not always smooth, that is never a smooth mingling of skins and languages, not always, but is rather, as Paul Gilroy suggests, "fractal"—jagged, sticky, mixed, thick.[59] And which Wittig captures by way of stretched skins—the libidinal economies that are always unevenly distributive. In this way, I am oriented by all that pulls me in or pushes me out, by all that I may respond to, and from which I may learn what it means to be a body—this body I have inherited and which I carry, as I must, and that carries me.

In considering these questions of borders and skin, from the broader frames of history and discourse to the particular scenes of certain skins, I want to conclude by reflecting on notions and understandings of community, this being-in-common, for community is often what is negotiated or narrated by way of borders and bordering practices. From national communities to communities of migrants, from communities emerging by way of struggle to those that dominate others, community is posed here as a critical frame for building scenes of negotiation, for bridging the often divisive lines imposed by regimes of state power. In this way, community may be approached specifically as a skin-work, a politics of the flesh of things in order to get at the limits of bodies as being the basis for what bodies share. As Monique Wittig suggests, in following the skin one is immediately involved in community, a togetherness always already political. In this sense, might borders allow us to think community not as a construct of fortification, but rather that of porosity and transition? As that which assists and supports in negotiating the dramas and traumas of being a border? If acoustics contributes to the work of reorientation, acting upon scenes of intimacy as the sharing of skin, a *consensuality*, what understandings of community might be captured through its discourses and manifestations?

In his philosophical work, *The Inoperative Community*, Jean-Luc Nancy offers a number of compelling perspectives which can help in thinking about community and how it may relate to the notion of skin-work I'm posing.[60]

In contrast to the figure of "the subject," understood as an expression of self-sufficiency, as the articulation of "sovereignty" and a state of independence, Nancy emphasizes the concept of "singularity," which is understood to be less a container of interiority, of self-consciousness, and rather, an exposure to others. Singularity, instead, as a finitude, a limit. "The singular being appears to other singular beings; it is communicated to them in the singular. It is a contact, it is a contagion: a touching, the transmission of a trembling at the edge of being."[61]

As such, a singular being is understood less as a self-sufficient holder of autonomy and more as a state of assemblage, a contagion, a tendency of transmission—as what is given, what can be given. "Singularity is the passion of being,"[62] and what it communicates is precisely itself. For Nancy, community is thus figured as the "sharing of singularities," a being-in-common shaped and fostered not as a form of work, a working toward completion, but rather through the qualities and experiences of passion, of ecstasy—a transmission, a trembling at the edge of being. Community, in other words, is emphasized not as a project of "communion," of what may restore a sense of completion (the narrative of fusion); rather, community as the unworking of being: community as what undoes being-as-infinite. Rather than community being defined by work, and the project of identification, of gravitating around a central identity, a narrative of nationhood, for example, community is the undoing of communion, figuring what I may additionally term a *composition of fragments*. This leads Nancy to the notion of the "inoperative," and the ways in which community specifically unworks being, staggers it according to a state of interruption. In short, community is a continual *propagation*, a movement that never fulfills itself. Rather, community is what fundamentally interrupts, is interruptive, straining any narrative of self-possession. This is precisely the generative dynamic of community: to remain on the borders of itself, as contagion, contact.

Echoing Bracha Ettinger's more psychoanalytic account, community emerges as what ultimately unsettles subjectivity, or rather, as what defines one's singularity—this *I* that is always already incomplete, in transition. Community is therefore not the overcoming of being incomplete; instead, it allows for a doing together, a sharing of singularity which is always already there—which is ontological. Importantly, Nancy moves to challenge the notion of "the lost community"—community understood as a narrative of a past unity, as having once been a coherent whole and to which one works at reclaiming, through any number of actions, or what Nancy calls "myth," whether of nationality or heritage. In contrast, community is a "literature," a writing whose story is never finished; "communication" for Nancy is not on the side of the bond, of the social bond. Within the framework of community, communication is rather that which exposes singularity to an outside: a passion, contagion, a writing, which may echo Ettinger's notion of "relations-without-relating."

Nancy's understanding of community may allow us to return to Monique Wittig, to read her queer writing along the lines of an inoperative community. Formed around the rapturous tearing of the body, an interruptive erotic state of being-in-touch, the lesbian body is narrated as an emergence of community: a queer, feminist community where the "I" is always already exposed, *ecstatic*, and which Wittig identifies throughout her book with a forward slash: the *I* [*J/e*] that contends with the violence she identifies as always already in (masculine, dominant) language (underscoring the gendered structure of French, where the masculine acts as a universal or abstract subject under which the feminine gains definition). Writing, for Wittig, thus works at breaking language, as the symbolic force that names (gender) identity and the communities we may come to form or desire. The inability to write "I" [*Je*] without subscribing to a patriarchal order is taken on as a literary act. Wittig confronts the task of writing with an "alien language" as she terms it; this language which is never *her* language, but one in which "feminine persons are complementary to the masculine."[63] Such an impossible task leads to the intervention of the forward slash (also in "m/y") that proliferates throughout *The Lesbian Body*. Such a typographic gesture figures violence on the page, unworking the symbolic power of language by way of a body in parts, an ecstatic literature. As Wittig states, the forward slash confers an excess to the "I" [*J/e*], exalting it to a state of visceral intensity: this *I* that cannot be held, that unfolds and mobilizes itself, communicating itself: as a passion, a contagion, a transmission.[64] From the emergent togethering of the lesbian body to the war waged by Amazonian women in *Les Guérillères*, Wittig's writing is a literature of and for community, as specifically aimed at the unworking of (patriarchal) capture and the heterosexual contract founded on masculine completion. A writing, instead, that plays at myth, that enjoys itself as it "lesbianizes" the idea of a foundational narrative—as it writes its own world.

In reading Wittig in this way, ecstasy emerges as the condition of and for community, as a sharing of singularities—a trembling at the edge of being—an inoperative unworking of (patriarchal) Being, a womb-narrative, and which we might begin to hear and feel as the spread and diffusion of a certain cacophony, a certain reverberation and vibration whose energetic force may give to the making of a borderspace.

> M/y fingers sink into the orifices in your back your loins, your fingers are inserted into the holes in m/y neck m/y cranium. In the end a tempest arrives, it rushes right through us, scattering the muscles. First *I* hear your cries, then *I* hear myself cry out as you do, there is a bellowing of sirens, they reverberate within the gaping tunnels on either side of our two bodies which constitute a single organism pervaded by vibrations quivering full of its own currents, is it not so m/y dearest?[65]

A quivering noise that may be heard as part of the work of reorientation—whether on the imaginary islands, or in the gay bars in Ljubljana. Or further, in the movement of the squares in Greece where, as the urban theorist Stavros Stavrides has sought to describe, community emerges precisely as a cooperative *movement*. Turning from Wittig's literary act and toward Stavrides' reflections on community, I may identify a concrete instance of community and its inoperative expressions, one that articulates and mobilizes an ecstatic remaking of public space. From within the movement of the squares, and the creative resistances found in Syntagma Square starting in 2011, "communities in movement," as Stavrides suggests, emerge as the diverse coming together of people in support of shared resources, cooperative practices, and the making of radical care networks: a social economy founded on a distributed sense of shared responsibility.[66]

Following the occupation of the squares, and the subsequent new articulations of urban life found in Athens in recent years, Stavrides underscores how communities in movement are fundamentally porous, articulating connections and linkages across the city that, in the case of the austerity crisis, give way to manifestations of "common space": a decentralized construct collaboratively shared, maintained, and that gives way to a profound sense of collective power. By way of the articulations and actions of communities in movement, people attend to each other in greater ways, and in doing so, capture the city as a social and material resource, a frontier of invention sustained through practices of mutuality. Such actions were brought forward within the occupations, and which manifested in a range of initiatives, from neighborhood medical clinics to academies held in the parks. Within the dynamics of commoning and common space, along with an ethos of mutual care, the operations of work are driven by cooperation, and the shared passions that mostly sidestep the capitalist drive toward profit. Consequently, the city takes on another dynamic and form. In tension with the structures defining the city as private or public, as market-oriented or state administrated, communities in movement contribute to the making of a "common city," one constituted less by the labors shaped by neoliberal economy, or by the channels and flows regulated by state administration; rather, common space is articulated more by the passions of cooperation that pass among a plurality of people in the making and collective responsibility of common life.[67]

Importantly, Stavrides notes that communities in movement and common space are reliant upon a dynamic of "thresholds." As a mechanism, thresholds define points of contact, a liminality that is not so much a disintegration or loss of form, of singularity or even identity, but rather, one that provides continual exposure. Thresholds turn borders into "turning points" rather than "check points" as Stavrides suggests. Upon a turning point—or what I may also underscore as a "transition point"—one confronts all sorts of risks as well as potentialities, for the threshold is "always being crossed."[68] In this perennial state of crossing, borders are not so much breached as they are activated—"constituted as a space of potentiality."[69]

Stavrides' notion of thresholds, and the figuring of a common city, can lend to understanding the bordering work I've been mapping here—one that does not proceed by way of policing, but rather by facilitating. This is to recognize how communities in movement are not without their tensions and contradictions. As Stavrides highlights, these become operative: "Discrepancies, ambiguities, and contradictions are necessary ingredients of a potential community in action, a community of different people who remain different but recognize themselves as co-producers of a common space in-the-making."[70]

His analysis of the movement of the squares, and the expressions of community arising out of the austerity crisis, can be seen to extend Nancy's inoperative community, as the passionate sharing of singularities. From another perspective, communities in movement and the social economies of mutual care may capture Wittig's lesbian community, moving the erotic strength of consensuality into the life of the city—as a revolution waged against the patriarchal profiteering that encloses and extracts. Here, the thresholds Stavrides finds in Athens act as physical and psychological mechanisms by which forms of bordering work may take place—a *tough love* specifically aimed at unworking the hard border in favor of "passage, contact and negotiation with otherness."[71] In this sense, community can be thought less as the fortification of borders, and more as a critical touching, a common sense, a transmission that deploys an erotic strength in writing the common city.

Understandings of community are therefore tensed by different positions and politics: by detention practices that fortify, that govern by way of all types of exclusionary techniques and strategies; or by the commoning endeavors that seek to threshold the city into co-creative, directly democratic, and consensual movements; by capture, extraction, or by opening, touching: by communion, or by community; by love for the one, or by love for each other.

* * *

Throughout these chapters, I've been attempting to capture ways of understanding the formation of subjectivity from an acoustic position, and further, to reflect on what consequences this might have for approaching living life together. From practices of compositioning and the formation of poetic ecologies of resonance mapped in the previous chapters, to the skin-work and borderspaces that emerge as defining features of what it means to be a body, I'm drawing out acoustics as a passage toward thinking of community specifically by way of (con)sensuality. From the inoperative community, and the porosity and jointedness defining its expressions, the rhythms that punctuate the drive toward collective movements, or the vibrational gestures that work at ecologies of feeling, at shared listening, what I'm searching for are the means to query the acoustic crafting I sense as being key to supporting diverse manners of living.

Although experiences of sound and listening may deepen relations with others, as the means by which to foster sharing, and which extends capacities for empathy as well as argument, they do so by also exposing oneself. Such exposure is understood not so much as a form of display, a presentation, but rather an agitation, animation, a *thresholding*: an *opening up*, *drawing one out*, where listening carries one forward while drawing others into oneself, and which may hinge language upon an axis of shareability, betraying or subverting the myth of the lost community and the desire for completion. In this regard, I am led to ask: How is it possible to remain myself as I listen? Am I not always already displaced, touched, unworked according to the acoustic figuring of being-in-relation, a being-in-touch that listening instigates and supports? And might a feminist listening, a queer listening, a *skin listening* draw out a politics of recognition and of location as what requires a generative, considered acoustic, a conscious touch?

The conceptual framework I'm working through here contributes to understandings of acoustic justice by specifically addressing the body as situated: a corporeality always already linked or strained by a politics of recognition and of location, where skin plays a defining role. By way of the sensualities and consensualities of being in one's skin, acoustic justice may engage the temporal and spatial horizons of the globalized everyday, and the crafting of the sonorous envelope as necessarily a tensed and stretched auditorium. Within such primary envelopes, from the rooms and streets that resound with so many voices to the streaming networks of digital media orienting us in numerous ways, sound and listening are figured within a greater economy of sensation, affection, and the intensities of touch that dramatically shape us, making of the skin a zone of contact, passion, and even debate. From such conditions, acts of orientation and reorientation impact onto the positionality of the body, seeking to define or capture routes for navigating through particular social and institutional environments: to find points of connection and conjunction, or to break with others. In this regard, recognition, and the possibility of just relations, is impressed upon the skin, finding traction and meaning by way of a body-politic. As such, the micropolitical economies of desire, which underpin the ways in which we come to touch and be touched by that which we long for, impact onto the ways in which social equality plays out. This takes on a more pronounced form, a more macropolitical form, when it comes to migration and the desire, if not the necessity, to move through the world, across scenes and contexts shaped by a range of conflicts and ambitions. Here, the status of the body is placed within a geopolitical system of government, whose acts of tracking and profiling impact a larger understanding of rights and recognition.

Following such perspectives, acoustic justice may capture the urgency around questions of orientation and reorientation: how bodies are always negotiating the forces that define ways of moving, and of being in contact. From queering the straight lines of heteronormative society in support of other desires, to assisting in movements over or against particular regimes of detention, acoustics is underscored as part of the support structures we may capture or create, attune with or interrupt. The psychic and sonorous envelopes that help in mediating our sense of being in place may be tuned or composed in an effort to recalibrate emotional and political life—thereby accentuating the border, the edge of oneself, as a threshold, a space, even a bar or auditorium. Through the dynamics of touch, borders materialize the ways in which contact is extremely demanding, and also the very thing that sustains us. Such conditions, as Stavrides suggests, may form the basis for fostering the power of people in galvanizing a collective crafting of life on the streets, lending to the movement of communities a deeply consensual force—a consensuality to be deployed precisely on the edges where community lives, transmitting itself.

4

Deaf Attention

Peripheral Visions, Spatial Meanings, Sensory Politics

Understandings of listening oscillate across an array of perspectives and connotations, giving way to a range of ethical, ecological, social, and personal positions. As an activity naturally tied to hearing, listening is equally figurative, metaphoric, and articulated as a means by which to deepen sensitivity for the shaping of relationships. As such, listening comes to extend and elaborate one's sense for the body, figuring modes of understanding that align or mediate exterior and interior, from the social worlds one inhabits and addresses to the flows of bodily life. Listening is often positioned as what may foster critical and creative attention, a vocabulary for capturing rather holistic experiences of embodied knowing; not only sound as it touches the ear, stimulating the cochlea, but more: listening as it disperses across the skin and the eyes, as it extends hearing to arrive on the tongue, within speech and the nurturing of conversation and gesture, and further, by way of the organs, the gut which tells us many things, that gives way to the excited, nervous, and passionate fluttering to which one attunes by way of an inner attentiveness, an inner listening. As the choreographer Lilia Mestre suggests, listening is what may productively confuse distinctions between conscious and unconscious being.[1] This finds a compelling elaboration in what Gemma Corradi Fiumara underscores as listening's propensity to supply human subjectivity with the capacity to "reconnect" biology with philosophy: to recognize the profound ways in which human "animality" (this corporeality) participates in our most rational views.[2]

Listening may be highlighted as a gesture aimed at fostering an elaborated sense of relationality and seems particularly adept at extending the limits of singularity. While listening may certainly be utilized in acts of surveillance, in regulating or arresting the circulation of voices and views, it is also often what may nurture sensitivity for the importance of such circulations, in which to listen is to work at greater understanding. To listen, in this sense, is to open a time and space for accommodations, generosity, and care as well as dialogue and disagreement. In this regard, listening is not restricted to the ear per se—rather, listening shows itself in the turning of the gaze, in the extension of the hand as it reaches, in the speech one gives and returns, the foot as it taps, as one smiles, or laughs, all these gestures one makes flex the coded contexts of social life. Listening, then, is the expressive means of *attention* itself: an attentiveness that engages with things, that moves across verbal and nonverbal sharing, and that cultivates a greater view, contributing to what Anna Lowenhaupt Tsing terms "the art of noticing."[3]

In this regard, listening emerges as a vocabulary—a power—for not only capturing a range of sensual, thoughtful, and relational perspectives; rather, listening is positioned within the thick of relations, precisely as what may work at aligning sense and knowing, feeling and understanding, biology and being, as what may bind body and thought, aligning touch and word, emotion and action. To listen to others is therefore not only about aural attentiveness or verbal speech; it is also about a greater economy of attention and process of recognition—a cognitive act; it is about proximity, intimacy, as well as distancing and refusal—from the small agitations of bodily sensation to the social and political movements that entail a communal perspective, a global argument, and understanding.

The movements of attention fostered or staged by way of listening point to a greater synesthetic and multimodal capacity: a perception that involves the whole body and, in doing so, figures a sensory politics. In this regard, listening emerges as a process of continuous learning and unlearning, a capacity for reflection that equally questions the ways in which attention is directed toward particular voices and things, how it is framed within a larger mechanism of social recognition and knowledge: for example, how attention is often directed toward hearing the human voice solely as that which is meaningful. Or, as Roshanak Kheshti poses, listening as it is shaped by "regimes of aurality," which impact onto the formation and figuring of subjectivity, contouring the order of the senses with a cultural politics.[4]

In this sense, acoustic justice may set out to question regimes of aurality, economies of attention, and a sensory politics so as to bring into question the ways in which normative understandings of ableness and sense perception are culturally re-enforced. This has at its base a demand to articulate a right to listen not only in terms of legal hearings, or the importance of the right to reply as central to public discourse and media practices, but equally within any number of public, private, institutional, and noninstitutional spheres. The right to listen is posed to also emphasize the importance of hearing beyond a dominant tonality or soundscape, as well as a dominant bodily schema often defining the "able listener," so as to attend to the diverse doing that is life. This is to recognize the degrees to which listening's capacity to foster empathy, compassion, and care, to productively confuse the lines separating conscious and unconscious life, bodily sensing and thinking are always situated within or around genealogies of injury, prejudice, and pain. Regimes of aurality, in underpinning a distribution of the heard, do so by amplifying particular sonic matters over others, giving traction to particular economies: social and sensual, financial and libidinal. In this regard, if the right to listen is to work at forms of justice, intervening within such economies, it becomes essential to consider the often entrenched frames that come to delimit (and capitalize on) listening's reach.

If listening, as I'm suggesting, is never only about hearing sound, but rather is located within a range of philosophical, bodily, social, and judicial registers, acoustic justice may offer a platform for contending with the work of listening, as the work it must do in wielding attention as a creative and critical power, as an art of noticing. Acoustic justice considers the ways in which the movements and meaningfulness of listening are deeply shaped and informed by a range of experiences, histories, and legacies of both justice and injustice. Such aspects are also greatly amplified by way of public discourses, media channels, educational institutions and their transmissions, which impact onto what we come to hear and to know. At times, the right to listen then might require that one *wrongly* listens, so as to challenge, circumvent, or pirate systems of power; to hear against the grain and the sonic matters that often bind us to particular views. Acoustic performativity, and the gestures and practices that may reorient the economies of attention in acts of not only sonic warfare but also acoustic care, becomes central to contend with such distributions and imaginaries of the heard. And which may work at manifesting what Fiumara further calls the "responsibility of listening."[5]

Péter Molnár, a poet and human rights activist based in Budapest, highlights how discriminatory government practices against Roma communities in Hungary have escalated, particularly following the parliamentary elections in 2010, which saw the inclusion of Jobbik, a far-right anti-Roma party. Mostly living in urban peripheries, and within impoverished areas throughout the country, Molnár suggests that attempts to remove the Roma from Hungary are enacted not only through direct attacks but also through "acoustic separation." Such acoustic separation, Molnár argues, allows for a type of erasure that equally forecloses possibilities of exchange and sharing across social groups, a separation aimed at undercutting the chance for telling and hearing life stories, especially those stemming from the realities of oppression Roma communities continually experience.[6] In this instance, the inability to narrate oneself in the face of such discriminatory situations lends to erasure and the perpetuation of prejudicial views. Acoustic separation, in this sense, limits the articulation of a certain visibility by cutting against the potential circulations of voice and the narratives one might tell and hear—and which significantly contribute to the shaping of *demography*: this "writing of the people," not only by way of statistics, but equally as a writing that captures the complex story of people, allowing them to hear and to read of themselves in an enriching manner.

Following Molnár, acoustics comes to impact the discriminatory shaping of social misrecognition, influencing the ability to foster social, political, and legal access and participation by way of the circulation of voices and views—to feel oneself as a subject, as recognized. If listening is about a greater economy of attention, an affective channel by which to touch and be touched, and that supports the dialogical and public flows of understanding and acknowledgment, acoustics is part of what one works at in order to shape or reshape a balance of power. Acoustics, as the distribution of the heard, is operative within those very processes underpinning an ethics or an imaginary of equality and respect, as well as the institutional structures whose task it is to support people.

Although I've been edging toward the limits of sound and listening throughout the previous chapters, arguing for acoustics as the basis by which to mobilize a greater view onto questions of justice, recognition, orientation, and community, I want to move further, to engage this web of complexity listening seems to suggest. To do so, it becomes necessary not only to consider the architectures of social and political environments, the ecologies of resonance as well as the work of borders that align or misalign skin with national territory, but to also firmly put into question the often dominating legacy of "audism"—what Raymond Antrobus calls "audio supremacy"[7] which, as Deaf studies and culture have shown, privileges the hearing of sound, especially verbal speech over other forms of communication and by extension, diverse forms of intelligence and ability.

Audism, as "a system of advantage based on hearing ability,"[8] is founded upon a larger metaphysical and moral understanding that centers humankind around verbal speech and the concomitant capacity to hear. Audism, accordingly, must be highlighted as producing a set of vocabularies and knowledge regimes, languages and pedagogies—"a regime of aurality"—defined by the capacity to hear sound and that denigrates or marginalizes other ways of accessing sound, that restrain listening (and language) precisely to the operations of the ideal ear. As Harlan Lane has suggested, by following Deaf culture one may learn of other ways of speaking and hearing that lead beyond phonocentrism; where signing, as a language, figures particular forms of perception, voice, poetics, and knowledge making. And where often dominant conceptualizations of sound, silence, and hearing may be upset, extended by way of D/deaf experience and expression—to give way to other life stories and other modes of narrating.[9]

As David Wright suggests in his personal account, deafness is more a "different" experience of hearing rather than its loss or absence. As he adamantly states: "there is no such thing as absolute deafness."[10] From music to footsteps, car engines to drills, low-flying airplanes to the roaring of crowds, Wright suggests there is nothing much he misses from the auditory experience of the urban milieu.[11] Rather, he hears these events by way of "touch-hearing," those physical, vibratory energies that pass from the world across his body, giving way to a form of hearing that shifts understanding deafness as being without sound. The Deaf poet and performance artist Aaron Williamson gives further insight: "But the imagery of silence is also a problem for deaf people. It's not exactly a derogatory perception, but like I said, it's a romantic hyperbole, to imagine deafness to be like luxuriating in a blissfully silent warm bath all the time. Deafness can in fact be a very loud world to live in."[12]

By shifting perspectives on how hearing takes place, and the ways in which diversely abled bodies perceive the world and develop ways of knowing and communicating, D/deafness allows for an important critical rethinking. What becomes of Sound studies from the position of D/deafness in which sound is reoriented toward that of embodied sensing, a "listening through the bone" as the poet Willy Conley states?[13] What theories of listening might emerge by engaging the greater economy of attention one's body is always already involved in, and onto which D/deafness offers a specific perspective? Shifting from an audist, or oralist, view of speech, what becomes of understandings of agency, and the ethical and legal struggles over the freedom of speech? And further, what might D/deafness offer in terms of approaching understandings of acoustics, as well as acoustic justice?

To follow these lines of inquiry, my approach is both argumentative and speculative. It feels important to follow listening as it is articulated and understood by way of D/deafness, leading as well to particular understandings of embodiment, language, and voice. I take guidance from a range of scholars working in Deaf studies, as well as through the works of D/deaf artists, performers, and poets whose creative approaches do much to challenge the hearing world, as well as various theoretical positions with regard to Sound studies. Given that my personal relationship to D/deafness is limited, I have attempted to pinpoint a number of key perspectives gleaned by way of readings into D/deafness, and primarily in relation to the visual-gestural languages of Sign. These perspectives include peripheral attention (and the peripheral as a concept), listening through the body (embodied listening and a politics of touch), and gestural voice (language as movement, as motility). These perspectives are explored not only by readings into D/deaf experience, and the perceptual, linguistic, or philosophical shape of Deaf culture, but additionally by considering what Paddy Ladd underscores as the "colonialism" at work within histories of deafness: the discriminations that abound within legacies of deaf education, institutional and governmental policies that have often relegated Sign to a state of non-language, and the various instances in which the deaf were (and continue to be) cast as "savages."[14] Deafness thus triggers an important and compelling social and acoustic justice framework for considering struggles over recognition, of what it means to speak and be heard, and the (post)colonial politics continuing today as the Deaf nation seeks new forms of articulation and solidarity.[15] This is not only about contending with the social currencies of sound and hearing, but also to draw out questions around medical and technological approaches to sound and hearing. For instance, cochlear implant (CI) technologies and medical procedures instigate complex challenges and discussions around diversity and the ethics of medical practice. From such a perspective, a notion of "bioacoustics" may be captured, underscoring not only the acoustic ecologies of soundscape life, but equally how understandings of sound and hearing play out in terms of bioengineering and medical approaches toward life.

To unpack issues of D/deafness, and its impact onto an acoustic justice model, questions of peripheral attention, listening through and by way of the body, and notions of gesture and voice become paths by which to reflect upon D/deaf experience and knowledge. And from which understandings of acoustics, and a subsequent politics of listening, of attention and recognition, gain critical input. In his introduction to *Open Your Eyes: Deaf Studies Talking*, H-Dirksen L. Bauman highlights how signing greatly affects the perception of D/deaf individuals, of D/deaf bodies and minds, specifically by activating one's "peripheral vision": signing brings into play a dynamic integration of visual, gestural, and spatial capacities, where the hands work at figuring language in the air.[16] Signing proceeds by shaping language as a spatial form, unfolding as a gestural action that is not only visual, as what requires one to look, but that additionally proceeds by way of positioning the hands in space, as a form of enactment. "Whereas hearing people depend on sound for information along the outskirts of their sight, Deaf people's vision pushes the boundaries of the peripheral, affording them the ability to entertain nuanced sensory input across a wide field of vision."[17]

The periphery, and peripheral processing of information, interestingly becomes central to reflecting upon D/deaf perception and D/deaf thinking. Through signing, peripheries become deeply participant as the gaze is directed at a constantly moving arena; looking performs a type of "visual embrace" that actively takes in a wider field of information, and that assists in establishing and maintaining communicativeness.[18] The impact of such perceptual skill lends to a deeper capacity for being able to "focus," for bringing attention to the act of communicating, as one's gaze must continually attend to others—for "to look" is often also to focus. Looking becomes adept at staying close to the flow and punctuations of exchange, to what is happening in front and around, which, in the case of signing, is quite different from verbal speech, where listening often drifts, grows distracted (where one can look away even while having a conversation). In contrast, D/deaf attention and communication is concentrated and highly animated.

The visual embrace, enacted by way of the intensities of looking and signing, further supports a heightened sense of intimacy, for the eyes and the hands are deeply sensual, expressing both tenderness and engagement, a touching, and as such they help produce feelings of affection and concern; to constantly address others by looking intently, and by continually touching by way of the hands, signing brings into play the conditions for a heightened sense of closeness. It might be construed that such capacities and experiences, which draw in the periphery and give concentration to a physicality of exchange, lend significantly to developing a larger, global view onto the world while equally fostering a more pronounced bonding with others.[19]

As a consequence, D/deaf individuals develop an acute capacity for comprehending a spatialization of meaning. In signing, language is much more bodily and performative, and much more spatial than in verbal speech. While gesture, as David McNeill poignantly reveals, is central to speech, and the social performances of verbal communication, with signing, gesture becomes the very means of "speaking."[20] This develops on at least two important levels, from the direct manual construction of a sign to the larger engagement with others—from making sign to maintaining social relationships. Sign language (e.g., American Sign Language (ASL)) is defined by way of five "phonemic" parameters including handshape, movement, palm orientation, location, and nonmanual markers, such as facial expressions or bodily shifting. These are essential to distinguishing one sign or expression from another, and integrate an important spatial component. It is not only a question of the shape of the hand when signing, but also the relation of both hands to each other, and the orientation and placement of the hands in space. For instance, in sentence construction one describes or classifies verbs by way of location and movement of the hand (for example, if one is describing the act of walking down stairs, the hand signs and descends at the same time). This is further extended when entering into conversation. One's look must monitor a greater spatial arena of signing, as one follows an often simultaneous procession of movements and expressions, figuring gesture as a social choreography: a shared "kinesthetic play."

Importantly, this does not stay fixed on the hands only, but incorporates a significant degree of facial and bodily expression. Benjamin Bahan, for example, demonstrates how facial movements and eye direction contribute greatly to D/deaf communication. "Eye behavior" in signing works at punctuating story telling, directing one's attention to shifts in narrative, highlighting spatial positioning of different "voices," and assisting in taking turns among a group of signers. As Bahan elaborates: "Eye movement may occur over a single word to convey specific meaning, appear in sentences to indicate the spatial position of the object, signal constituent boundaries, bring the addressee in and out of a story world, and/or play a role in turn-taking."[21] This array of eye behavior, and subsequent spatializing, often occurs within seconds, creating an extremely adept sense for conversation as spatial, visual, and bodily, and may also often include modulating the opening and closing of the eyelids—from squinting to opening wide, from eye-blinking to the closing of the lids, eye behavior forms an array of signals that punctuate and further inflect signing.

Questions of peripheries, spatialization, and the kinesthetic figuring of meaning and communication become a conceptual base from which to reflect upon a phenomenology of D/deaf experience, the primary work of perception, and also to speculate as to the spatial, temporal, bodily, linguistic, and relational meanings and capacities that open up within Deaf culture. Are there forms of peripheral, spatial, and kinesthetic thinking that emerge and take hold within D/deaf experience, and if so, what might such capacities suggest or engender in terms of articulations of community and of attending to the world? What type of impact might social, linguistic, and corporeal conceptualizations of D/deafness have for thinking acoustics as a form of practice and justice work?

In shifting from speech to sign, listening also migrates to that of seeing and feeling, whether in the elaborated eye behavior that punctuates signing, or in the ways in which the D/deaf hear the world—by sensing vibrations and

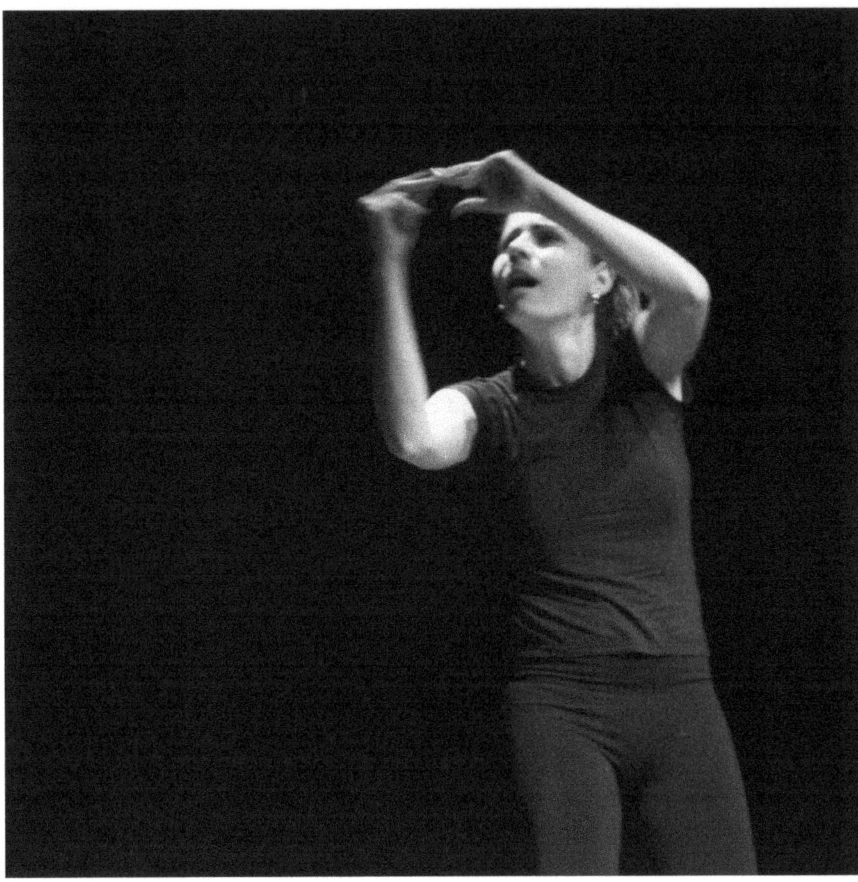

Figure 4.1 Mariana Santos, in "Surdo, Logo Existo!" ("Deaf, therefore I exist!"), Deaf theater work on the history of Deaf struggle, Teatro Guaíra, Curitiba, September 2019. Photo: Gilson Camargo. The performer is asking herself: What is happening? Is it true that people will not be allowed to communicate with their hands? The performer is living the historical moment at the Milan Conference of 1880 in which Sign was officially prohibited. The hands refer to the action of two people speaking to each other.

movements, or by following the hearing of others (often D/deaf individuals perceive that something is aurally happening by seeing the reactions of hearing people). Listening, in this regard, is not absent, nor is sound; rather, the acoustic arenas that shape the experiences of not only hearing, but equally that of attention and sensing are also captured by D/deaf experience, a D/deaf positionality, reorienting a particularly audist view toward that of embodied, visual perception and ability.

In addition to the importance of peripheral attention, as well as D/deaf listening, understandings of D/deaf voice shift from that of verbal articulation to the gestural. The "gestural voice" displaces understandings of voice as being bound to the vocal organs, and instead moves to that of the arms, the dexterity of hands and fingers, facial expressivity and the overall movements of the body. Voice and language, in this regard, are dislodged from an overarching phonocentric, oral framework and are given articulation through the eloquence of Sign.

The groundbreaking work of linguist William Stokoe in the 1960s was to significantly upset understandings of Sign as "pantomime," as a "primitive" form of pointing. In contrast, Stokoe showed how sign languages are sophisticated linguistic systems that carry not only communicative ability, but enable the articulation of conceptual thinking and the expression of culture. Stokoe, in fact, suggests a primary link between gesture and language, highlighting the ways in which processes of naming, and giving meaning to the world, are bound to a question of gesture.[22] Stokoe's work has had an extraordinary impact on Deaf culture (as well as the science of linguistics). By revealing Sign as a recognizably complex language system, Stokoe upset entrenched views of deafness as a disability (often equated with a lack of cognitive ability), while also greatly emboldening Deaf culture. Consequently, understandings of "having a voice" also radically shift.

As such, what further consequences does this have for understanding language as often being bound to phonetics? And further, what becomes of knowledge if one shifts away from written language, as a set of signs often grounded in verbal speech and notions of a hearing community? How does gestural voice, and the kinesthetic, visual, and spatial dynamics of signing impact onto writing and the genealogies of knowledge, especially when "writing" is inscribed or published by way of video recording?[23] And finally, what becomes of a politics of recognition within a D/deaf framework and by way of what Paddy Ladd terms "Deafhood"?[24]

To follow such questioning, I continue by elaborating some of the key points highlighted earlier. I focus on teasing out, through a diagrammatic and speculative approach, what D/deafness offers in terms of thinking acoustics as a question of recognition and politics, community and justice—how understandings of "having a voice" require continual thinking. In speaking about deafness and Deaf culture, I recognize that there are a plurality of Deaf positions, and that within the Deaf community there are multiple ways of experiencing, understanding, and articulating a Deaf identity. In this regard, I realize my own approach may generalize D/deaf experience. Given this, I also try and stay close to D/deafness (by way of readings and reflections on Deaf studies scholarship and Deaf cultural practices), searching for productive views onto the topic that can also intersect with other intellectual fields in and around Sound studies. It is definitely not my intention to "exoticize" D/deafness, but rather to follow existing research and cultural work, and in doing so, to learn from their example. And more, to reflect on what such examples and perspectives suggest in terms of approaching acoustics and the issue of acoustic justice. By doing so, I intend to bring into question how understandings of sound and listening, and the "responsibility of listening" Fiumara calls for, may gain from engaging with D/deafness, both as a physical condition and as a cultural framework.[25]

The Gaze That Attends, Communicates

Where does language begin? At what point does communicability emerge as a gestural vocabulary to open a path to language? A beginning that is the opening of relation: in that moment, or scene, of being one in front of another, but even before, the moment of being held in the arms of a parent or guardian (this first home). At this moment, language begins as an experience of sensation; communicability—as the reaching toward, as responsiveness—might be said to be *all over the body*: in the touch that enwraps a child, that passes by way of so many caresses, emerging within this scene of attention, of attentiveness. Attention, as what Jan Masschelein highlights, is the ability to be present, an art of "making present" that also works at making "experience possible."[26] Attention is already in the hands, along the skin, in the gestures of care: a gestural vocabulary by which the senses learn of themselves and their reach, their capacity and responsibility (what takes shape, there in this scene of holding and responsiveness). From hands to feet, lips to hair, the belly and the bum, and further, in the gaze that attends, that turns and is continually passed between child and parent or guardian. The gaze that binds one to another, that is the binding by which many other lessons and leanings follow (the moral and social laws of looking). The gaze is a communicative act, a distance bridged, sustained, flexed by way of the look—the eyes that set the scene for acknowledgment, for love, even from across the room, your gaze gives reassurance, it says: *I see you*, and therefore, *I am here for you*. I am attending to the relation that defines us (I care by way of the eyes).

Stephen N. Malloch highlights such communicative acts by way of "musicality," which allows for an analysis of the rich performative movements of the gaze as they establish types of motifs, figures, and pulses; forms of repetition, for example, of cycles shaped by passing back and forth a look, and by the penetrative, soft, intense, and endearing qualities of gazing, dreamily, intently, and which enables the "creation of balance." (A coming and going, appearing and disappearing that Freud also narrates in "Beyond the Pleasure Principle" by way of "fort-da" (gone-there) in the case of his baby grandson: the coming and going of the parent, the presence and absence of the mother, which the child is always susceptible to and which is turned into a game by the child throwing his toys and pulling them back with a string: a rhythmical conditioning shaping the ego and processes of communicability and attention; we might already find this in the smaller movements of the gaze, in the gesture made by the ongoing presence and absence of a parent's face and eyes, as they turn toward the child and then away: a musicality taking shape through the performative modulation of looking—the look that returns and gives way to an exclamation of joy: there![27])

Communicative musicality is posed by Malloch as a prelinguistic arena, giving way to "timings" and "spacings"—a flow that is always "balancing" (synchrony) itself by way of vocalizing, caressing, rocking and cradling, and through the turning of the gaze, all of which works at defining "coordinated companionship."[28] And further, what Malloch calls "narrative": the "creation of meaning" through shared activity with others[29] which, in this case, is captured by way of musical articulations (musicality as the making of a compositional expression: attunement jointly crafted, or tensions held between, in the touch of skin, in the eyes that follow and therefore support another, that greet or encourage, in the pulsing and frequency of "ocular phrases"). Such processes and expressions work at nurturing the capacity for attachment, giving a sense of security to the child which, as John Bowlby importantly found, shapes a greater psychic and emotional frame through which one learns how to navigate the comings and goings of future loves and losses.[30] In this regard, looking and the bonds it may foster unfold through a variety of small modulations and articulations, where eye-blinking, for example, takes place as a subtle gesture that may work at soothing, holding, or that may play at all sorts of games.

Following Malloch's extremely insightful research, what forms of communicative musicality might emerge by way of D/deaf vision, as a gaze attuned to the power of looking as well as a wider arena of peripheral life? May we speak of "peripheral communicability," one particularly adept at drawing connections between the unifying partnerships of oneself and another, along with a broader scene? A musicality shaped by the tangential or the marginal, and that creates narratives by way of the visual-gestural language of Sign—narratives of peripheral attention? What forms of coordinated companionship take shape by way of D/deaf attention, which sees a perceptual field brimming with meaning, with movement that is meaningful? And that is also made sensitive to the comings and goings of others, of those one loves and is loved by, figured within a spatial arena alive with grammar, hands, sightlines?

Following such questioning, a D/deaf communicative musicality within the crafting of the first house (that of parental love) relies intensely upon line of sight. This extends beyond the extremely intimate and proximate relation of parent to child. Often within D/deaf homes, for example, sightlines are accentuated if not required, capturing the ways in which relationships within the home are shaped by *seeing* one another. As Matthew Malzkuhn outlines, the "cultural customizations" of the home enacted by D/deaf individuals and families rework space in support of a D/deaf lifestyle. This often includes "the making of a common collective space; the taking down of interior walls or making new openings between rooms; and the placement of mirrors and lights in strategic locations to extend sensory awareness and maintain visual connection between family members and guests."[31]

To see is to already communicate, which may draw us back toward the primary scene of the parental look and the life of attachment it supports—the sightline that communicates so much, and that may be found in the articulations of D/deaf space: the architectures and social environments crafted by way of D/deaf orientation. For instance, it is natural for D/deaf individuals and groups to form a circle when congregating. This extends into the architectures of institutional spaces as well, for instance in the classroom, where it becomes essential for everyone to see each other. D/deaf space, in this regard, is mostly a space of collectivity and of communal support, an orientation explicitly shaped by way of turning toward one another. Following such orientations, the musicality of D/deaf communicativeness opens a view onto how visuality, and the capacities found by way of looking, is fundamental to the life of social bonds. We may consider to what degree such ocular intensities and musicalities are carried along, from that parental gaze into the making of a D/deaf world. The sightlines of D/deaf space are suggestive for the construction of a world shaped by the felt connections that sustain us as individuals.

Visual Listening (Movement Politics)

The turning of the gaze, the behavior of the eyes, the interplay between seeing and listening, are significantly emphasized in D/deaf experience not only by way of the peripheral gaze, and the meaningfulness of Sign, but also, in what David Wright terms (by way of Wordsworth) as "eye-music": the bustling, shimmering, and jarring movements found in the surrounding world that draw forth a sense of audibility, a hearing driven by seeing movement. The "visionary noises" Wright describes experiencing are suggestive for approaching the senses as being synesthetically linked.[32] Synesthesia is posed as a framework for appreciating the connectivity and "cross-talk" between the senses, which finds a certain manifestation in D/deaf perception and cognition. This may also be articulated as a form of survival, a strategy of navigating the hearing world by *hearing through other means*, inferring sound from the movements and actions of others, leading also to questions of neuroplasticity.[33] As Jane Dillehay notes, the auditory cortex of the brain of deaf individuals has been seen to take on other functions, for instance in support of peripheral vision.[34]

Seeing by way of D/deaf perception emerges equally as a form of listening: a path of sensing within an economy of attention, and underpinning the musicality of communicability. In turning the gaze, one equally says: *I give you my listening*. Thus, the profound sense of "presence" figured by way of the "power of attention"[35] is fostered through the particular movements and positioning of the senses: to be *here* is to quite literally direct one's senses to the immediate, confirming the presence of others which in turn lends confirmation back to oneself: a form of interdependency and intersubjectivity underpinning recognition and acknowledgment and which is given traction through "gestures" of attention.

Recognition, while figured within a schema often related to social and legal rights, is equally shaped by the more nuanced processes of "sense" and of perception. In this regard, I may emphasize sensing as both what one receives as well as gives, and that initiates an experience that nurtures (or outlines) a "moral" or "intersubjective grammar" of respect and esteem. Here, what Axel Honneth terms "integrated identity"[36] emerging by way of recognition is found not solely through institutional or social structures, but equally by way of the world of perception.

A politics of rights and recognition, which is often a struggle over "visibility" and of "having a voice," brings into play questions of sense and sensation. I may speak of "the sensation of perception," as the sense one has at being perceived—in inhabiting the world of perception—and which grants a certain existential shimmer: that as one looks out onto the world, the world looks back.[37] In this sense, the return of the gaze is often also the opening of a social arena—a punctuation of attention: the construction of a sightline—by which one is immediately involved in the life of others; an arena that one works at, where under the gaze of others one may also sense the degree to which one is regarded, seen. In this way, recognition also needs confirmation, expression, performance: a sensing and a sensuality.

Recognition, as Honneth outlines, is fundamentally predicated on the first intimacies experienced as an infant: the emergence of parental love (returning us to Bowlby's theory of attachments). Consequently, I may begin to underscore the communicative musicality Malloch proposes as a path for engaging in a politics of recognition; that the essential formation of attachments, and the emergence and maintenance of social bonds, is punctuated by the ways in which one gestures toward others, in the turning of attention enacted: the listening and the looking one gives or returns. In this regard, the "space of appearance" that Hannah Arendt outlines, as the space of politics, is fundamentally not only a situation of "being visible," of rights and recognition, but one given shape through the gestures that confirm or upset visibility: that give to the space of appearance a particular dynamic, a choreography, a musicality.[38] Here, Arendt's emphasis on "action and speech" as forming the basis for politics, for an exchange between political subjects, requires an extended sensual frame—a question of the body as a feeling and sensing singularity with particular abilities (physical, linguistic, social).

The gaze that turns toward others is quite often a gaze that acknowledges, that takes a step toward fostering a path of recognition (which, as Honneth argues, forms the basis for love, respect, esteem): a turning that inaugurates a movement, a rhythm, a coming and going, that punctuates one's capacity for self-realization: the gaining of a sense of definition.[39] Recognition, as being founded upon experiences of love, and which underpins understandings and expressions of equality and solidarity, suggests a sensitivity for the articulations of attention and its economies always already impacting individuation and social life. Here, "communicative musicality" may be utilized to position the primary experiences of attention, of being attended to and of returning attention, within a framework of recognition, suggesting that "action and speech" also define a sensory politics: that questions as to how one is situated within spaces of appearance, their sensory logics, and how those spaces emerge as articulations of power. A sensory politics then may be underscored as what also works at constructing or reorienting the physical and sensual coordinates of a space of appearance, and that impact on how action and speech are to be understood or, in fact, seen.

D/deaf recognition, following the dynamics of Sign as a visual-gestural language, along with the intensities of peripheral vision and the work of sightlines, may suggest specific types of musicality—actions and expressions that work at figuring, through a range of articulations, movements, gestures, and "customizations," D/deaf spaces of appearance, where *to be seen* is radically meaningful. This suggests an ongoing challenge for D/deaf recognition: the space of appearance, as Arendt suggests through her reflections on the Greek *polis*, is civilized by way of speech—by what Plato highlights, in the dialogue of *Cratylus*, as the capacity *to name* founded on verbalization. Man understood as the "speaking animal" places language firmly on the tongue (*lingua; language*); humankind thus emerges with the capacity to speak, to verbalize, and thereby, to act as an informed subject within a space of politics, where action and speech define the public realm. As H-Dirksen L. Bauman outlines: "Plato securely locks in place the tongue and mouth as the exclusive organs of language—the default mode of human being, thus casting the hands, head, and rest of the body into another nonlinguistic realm."[40]

One is subsequently recognized (by way of such overarching audist structures) in so far as verbal speech becomes performative, as it regulates the order of the senses (of the properly heard), becoming the base by which to act upon the world. To shift to the visual-gestural language of Sign (as well as to delve into the movements by which D/deaf attention "reads" and "scripts" the world: this synesthetic configuring of knowledge) is to reorder agency away from verbal articulation. The space of appearance for D/deaf politics, and for D/deaf agency, is instead an arena of gesture, sightlines, of visionary noise: a *movement politic* that puts recognition on the hands. "We use our bodies, our hands as a syllable."[41]

David Wright offers insight into how as a deaf person he experiences "hearing" by way of "movement," highlighting how "silence is not absence of sound but of movement."[42] In this regard, it may draw us back to the musicalities Malloch poses, driven by the fundamental dynamics of the look, touch, a tactile holding and humming, a movement of the body—a prelinguistic arena of communicability by which attachment is performed and punctuated. Musicality, in this regard, is not so much a question of hearing, or solely of bodily sensation; rather, it figures as a compositional (and synesthetic) frame allowing for the affective flows and social dynamics of interpersonal and inter-sensorial exchange to take shape—that is, a "politics" to flourish—a shaping, a *compositioning* that is the articulation of a certain worldview.

Listening with the eyes, with the hands, shaping language with the body, nurturing sociality through sightlines of attention, is suggestive for a rich perceptual and cognitive engagement with the world, and which may figure D/deaf agency as working by way of connections between the seemingly dissimilar: where visibility and appearance are housed within a larger frame of embodied knowing, and that firmly positions "action and speech" as one and the same.

Spatial Poetics: Picturing

Peripheries, as I've mentioned, come to participate more fully within D/deaf perception, along with the intensities of movement, giving way to a greater sense of spatiality and the figuring of sightlines. "We see then, in Sign, at every level—lexical, grammatical, syntactic—a linguistic use of space: a use that is amazingly complex, for much of what occurs linearly, sequentially, temporally in speech, becomes simultaneous, concurrent, multileveled in Sign."[43] As Oliver Sacks notes, sign languages generate an active relation to space (is already a question of space): from peripheral vision to meaningful movement, from bodily gesture to social choreographies, Sign places language in the air, on the fingers, across the chest, upon the eyes, and subsequently, draws into play a spatial dimensionality in which narrative proceeds through forms of "picturing."[44]

Questions of iconicity, as the relation of word to thing, as a primary drive underpinning language, are brought forward in signing, figuring the act of naming, of nominalization, by way of gesture. As such, signing suggests an answer to Plato's inquiry into the nature of naming, and the origins of language, by drawing voice back from the oral cavity and toward the hands, the body, the finger that points, that traces the shape of worldly things.[45] As William Stokoe indicates, "visible human movements are not merely sufficient for language but were absolutely necessary for making that first solid connection between sign and meaning."[46]

In contrast to the fixed sign of the written, Sign is always a shaping, a picturing, a performative embodiment and enactment that retains a direct relation to the things it addresses, describes, names, and narrates: it is always reaching and in touch with things, animating a world between the hands. This extends to that of "personifying" that which is described, for instance, through alternating roles: in signing one often "speaks" or gestures from the position of "characters" being discussed or spoken about. (This is elaborated in the concept of "visual vernacular" in Deaf performance and theater, a technique developed by Bernard Bragg to enable a performer to remain stationary while alternating roles.[47]) This draws out a pronounced sense for the interweaving of objective and subjective positions, whereby alternating views of the world are physically "taken on."

In this sense, one may speak of a D/deaf phenomenology to capture in what ways D/deaf subjectivity is bound, nurtured, and shaped by a relation to things by way of visual-gestural languages and the bodily schema of gestural voice: by a *movement politic*. A D/deaf phenomenology posits the world, and the figuring of the body, through a visual intelligence—as Oliver Sacks asks: "Does a new form of visual *intelligence* become possible?"[48] by way of D/deaf perception—which is also bound to a dynamic sensing of space: a spatiality under the hands by way of the look that directs, that *accents* expression. Signing draws forward a visual intelligence that is also spatially attuned, where the question of handshapes is additionally one of location.

While there is a rich history of spatial practices in the arts, of critical spatial thinking, of radical urbanism and questions as to what constitutes the production of space, and that specifically searches for openings onto architecture and the built environment in order to remind of the body and the ways in which space is emergent, is co-produced by way of movement, by way of a body—in relation to this range of important work, D/deaf phenomenology and D/deaf spatiality can be appreciated as *fundamentally* performative: the visual and the spatial intelligence underscoring Sign and D/deaf perception leads to a sense for a D/deaf production of space as always already bodily, participatory, social, literal.

In arguing for perception and embodied knowing as a primary position, a way of being in the world, Merleau-Ponty works at shifting cultural and social meanings by way of the body; phenomenology of perception, rather, aims for the experience of corporeality, as what may lead us out from under the "sediment of knowledge."[49] As such, Merleau-Ponty makes a distinction between "perceiving" and "defining," accentuating a tension between perception and meaning.[50] D/deaf phenomenology, in this regard, is a thick phenomenology in which perception is loaded with meaning, with language and knowledge. By way of Sign, D/deaf individuals are already moving in language—motility and understanding are deeply entwined. As Raymond Antrobus draws out in his poems, language is only ever *with* the body, where the state of physical being is never at odds with language—never an interruption onto language; rather, Sign carries language along, fixing perception and embodiment to a richly expressive world of meaning and knowledge.

Here, I may question the view that would emphasize language and body, sociality and sense experience, culture and corporeality as dialectical, as being at odds. Rather, to open critical and creative opportunities for rethinking philosophies of the body, especially those working at questions of language and power. For instance, what becomes of Julia Kristeva's distinction between "the symbolic" and "the semiotic"—the process of "subjectivization" figured by way of language as something in tension with the body: as that which never quite "fits" with the body as a sensual entity, and as such, comes to constitute a "processual" subject, a subject "on trial." What becomes of such tensions within a framework of Sign and D/deaf subjectivity? In what sense do the symbolic and the semiotic participate in the particular logic of signing? The "semiotization of the symbolic" Kristeva seeks to capture (as found in avant-garde poetry), as signifying practices rupturing the socio-symbolic order—that gives way to a "motility where all meaning is erased"[51]—seems to require other considerations when confronting the spatial and visual intelligence and poetics of Sign in which motility and meaning are fundamentally united and mutually supportive. In addition, what becomes of Michel de Certeau's notion of the "semiotic body" that scripts itself as it walks, where walking is cast as a gesture of appropriating the text of the city, thereby figuring a type of pedestrian rewriting of urban life—a body that introduces drifts, escape routes, an "antidiscipline" against the symbolic ordering of the city? De Certeau's theories of semiotic reordering is equally based on a model of language superimposed over the body, suggesting a linguistic construct of the body that also contains or is built upon a tension, a drama between the symbolic and what he terms "enunciation," leading to the view that "language is real only in the act of speaking."[52] In this sense, for de Certeau the "speech act" is always aligned with a potentiality of "ruse," of appropriation: where "tactics" are performed against the "strategies" of the symbolic, giving way to the crafting of detours—walking as a textual practice (a reading, a speech act).[53] Following de Certeau's schema of enunciation and ruse, what type of city emerges by way of Deafhood, by way of the sightlines of D/deaf space, and the D/deaf script articulated as always already bodily? If language is real for the Deaf in acts not of speaking but of moving, what forms of everyday practices become possible?

Expressions of "bodily scripting," along with the visual and spatial intelligence Sacks considers, are dynamically at play in D/deaf poetics and ASL poetry. As the poet Jim Cohn shows, the "visible poetry" of D/deaf poetics does much to transform poetic tradition, as understood by way of phonocentricism and notions of orality, shifting from the page and verbal speech to that of a performative construct held in the air, on the ground, across the senses: an act of "picturing" or "painting in the air" that articulates and fulfills the (modern, avant-garde) poetic drive toward "image" over "rhyme."[54] For the signing D/deaf, the image or the picture is not an abstraction, rather, it is something extremely "concrete"—where "every movement means something."[55] As W. J. T. Mitchell suggests:

> All of our standard preconceptions about language and the senses are brought into crisis by the emergence of gesture. The distinctions between word and image, between telling and showing, reading and seeing, writing and speaking, listening and watching, all seem to shimmer and vibrate in the presence of sign language.[56]

Picturing is the basis for a literature of gesture, where one "places" into space a set of words, narratives, personalities, articulating them by way of movement, a figuring—by way of configuration, location, motion. An arena of expression that is *corpo-real*, a body deeply involved in what it describes—that personifies, or rather, embodies; one speaks of actions by way of action, aligning the body with literature, with the *literary act*, which is no longer referencing the page, script, or text, but is, rather, in the world (reorienting notions of literary production by way of what Annette le Fort highlights as "the phonetic skin"[57])—a *soma-poiesis*.

As an example, I may highlight the ASL poem titled "Poetry" by Flying Words Project. In the poem, the Deaf poet Peter Cook describes poetry through an array of extremely evocative signed images, from rain falling through trees to explosions that seem to splinter the body, as an experience of rapture. In one section, he images poetry as a blazing sun, signing the sun through outstretched hands held up high. Following this, he moves toward describing a red falcon flying toward the sun, at which point the wings of the bird seem to collect the rays of the sun, carrying them by way of a sign that is part-sun and part-bird, with his fingers outstretched and hands open wide as they fly into the distance.[58]

D/deaf poetics thus poses a "textual theory" that fully expresses the "motility" Kristeva aims for, and de Certeau theorizes, as that which may not only interrupt the totalizing tendency of the symbolic (and the enclosure of the subject), but that equally unifies language with the hands, the body, the face, and the feet (as ASL poets often "tap out" poetic meter by way of the foot), pushing literature into space, onto the ground, into the sky, as a dance of meaning—where poetry is a restless and refined movement, a *soma-poiesis* harnessing many forms and

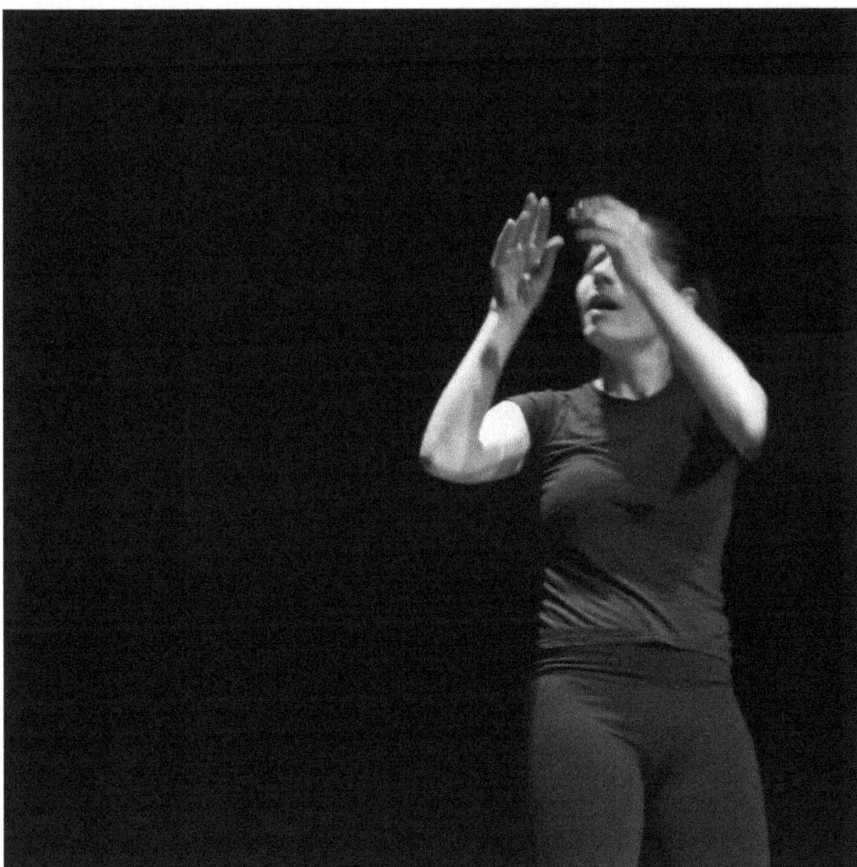

Figure 4.2 Mariana Santos, in "Surdo, Logo Existo!" ("Deaf, therefore I exist!"), Teatro Guaíra, Curitiba, September 2019. Photo: Gilson Camargo. The performer is performing the sign LIBRAS (Língua Brasileira de Sinais/Brazilian Sign Language) as the addressee of someone else.

forces. In this regard, might we speak of the musicalities found in the nurturing look, and the common life fostered through the sightlines of Deaf space, as the figuring of a certain unifying tendency—the ability to literally *carry* meaning, and to make meaningful the body as it lives linguistically? And that, in turn, makes language less a force to be reckoned with?

Gestural Voice

In her philosophical work, Adriana Cavarero argues for the voice as an essential sound, where the "vocalic" is never always already destined to speech and language, but rather first and foremost exposes the "uniqueness" of one's singularity (I speak and therefore expose my uniqueness: my voice captures my singularity, expresses it). In vocalizing, in resonating by way of voice, individuation essentially *stands out*, is made apparent. While Cavarero aims at undoing understandings of "the rational self" as being divorced from the primary vocalization of corporeality (*logos* as driven by articulated thought, a certain metaphysical authority), she proceeds by way of an overarching phonocentrism; as such, the "ontology of uniqueness" is expressed (if not founded) through sonority.

> In fact, the vocalic helps not only to think in ever more radical and corporeal terms the ontology of uniqueness, but it helps above all to conceive of politics in terms of a contextual relation, entrusted to speech, which does not appeal to territory or identity myths of community. The protagonist of this politics is a speaker who, leaving aside his or her belonging to this or that identity group, this or that language, communicates him- or herself first of all as voice. Thus, this speaker does not correspond to the universal, disembodied individual of the modern political fiction that relies on the communicative rationality of language. The reciprocal communication of voices implies the figure of a speaker who exposes first of all herself as a singular body. This is also true—as Levinas would say, and in a certain sense, Arendt too—for the face to face of those who, by looking at each other, expose their embodied uniqueness to the other's gaze. But as I have emphasized repeatedly, the acoustic sphere, unlike the visible sphere, is characterized by an essential adherence to Saying. And it is thus the most proper manifestation of that uniqueness and relationality that are announced as already inscribed criteria for the political essence of speech. It is therefore worth insisting on this decisive point. The voice is sound, not speech.[59]

What becomes of "uniqueness," and the sonorous essence of the political subject, when the voice is no longer sound but gesture? In what sense might individuation be retuned, even rethought, by complicating even further the philosophical and metaphysical views Cavarero seeks to challenge by widening "voice" toward that of looking, kinesthetic-picturing, and the dexterity of Sign—a "speech" centered around D/deafness? What type of "acoustic sphere"—as the sphere of uniqueness—emerges by listening to the Saying enacted by way of gestural voice, a listening enacted through the body, in the muscles, a "listening through the bone"?

David McNeill argues for understanding the relation between verbal speech and gesture as a form of "co-expressivity," highlighting the ways in which gestural movements participate in the articulation of thought. In doing so, he extends linguistics as being bound to worded meaning, suggesting instead that an "imagery-language dialectic" is at play in verbal speech (imagery being the "picturing" that gesture comes to create). As he states, "to make a gesture, then, is to iconically materialize a meaning in actional and spatial form."[60]

Importantly, gestures do not only assist in the work of language, aiding the articulations of semantic meaning; rather, gestures *are* language, constituting meaning in and of themselves. McNeill highlights the imagery-language dialectic then as co-expressive, underscoring the ways in which each (gesture and speech) destabilize *and* support the other: acts of communication become a processual work, a performative enactment of bringing thought into the world (thinking-worlding). Gesture and speech form a rich "dance" by which meaning, and the making of a social space (figured by the matters of language) emerges, is tensed and then resolved; where speech pushes gesture, drawing the body into play, and where gesture punctuates speech, leading thought, giving way to an organic "dialectic" continually at work (language as a movement: a *doing*).

This leads McNeill to engage with Merleau-Ponty's notion of language as a "house of being": as a question of inhabitation, of dwelling in the world, emphasizing language as a physical materialization which is continually shaped and constructed by way of linguistic acts. Although voice is often posed as being in tension with language (not only in Cavarero, or de Certeau, but also in the writings of Mladen Dolar in which voice is emphasized as being between the body and language: voice as a *paradoxical thing*—never strictly in one or the other[61]), the notion and expressions of gestural voice seem to suggest other views: where the body *is* language, and language is articulated by way of kinesthetic expression: a voice always already *figured*, bound to an expressive body.

Dorothy Miles' poem "Language for the Eye" (1972), originally presented on BBC TV Open Door program, captures the particular qualities of signed-voice, and the ways in which gesture and thought are interwoven.

> Hold a tree in the palm of your hand, or topple it with a crash.
> Sail a boat on finger waves, or sink it with a splash.
> From your fingertips see a frog leap,
> at a passing butterfly.
> The word becomes the picture in this language for the eye.
>
> Follow the sun from rise to set,
> or bounce it like a ball.
> Catch a fish in a fishing net,
> or swallow it, bones and all.
> Make traffic scurry, or airplanes fly,
> and people meet and part.
> The word becomes the action in this language of the heart.[62]

As a signed poem, "Language for the Eye" utilizes both English and British Sign Language and poignantly captures a sense for having language at the fingertips.

D/deaf voice may be understood less as a paradox, less an experience of "ventriloquism" as Dolar suggests (as what speaks *through* me by way of an "order of power and discourse," which is never fully myself). Rather, voice stays close to the body, *as* the body. It is embodiment itself: a body in action.[63] Such an approach puts into question dominant understandings of voice, along with understandings of D/deafness, upsetting oralist and audist positions by showing voice as uniquely articulated beyond a strictly acoustic sphere.

The Agency of Hands: Transformation Lines

There is a dynamic sense of simultaneity in signing, where the hands collaborate together, lending support from one to the other that work in parallel to the eyes, the face, the mouth: an elaborated "gesturology" that forces a reordering of linearity. Written language and verbal speech are profoundly linear, finding orientation along the "articulated path" of a "forward progression": writing, reading, and speaking follow a form of expressivity that takes its cue from the pen, the pressing out of letter by letter placed one after the other; in this regard, the line figures as a dominant principle and comes to direct patterns of thought, articulation, and feeling, setting relations along a linear course: of singular steps that work at temporal and spatial advancement. One first starts here, and then progresses, step by step, one after the other, figuring a past and a future upon a single line.

Verbal speech additionally sounds itself phonetically, along the course of a breath: this ribbon of speech that unfurls from the mouth. Yet, when turning to the hands, and the operations of Sign, the line is immediately multiple: one must proceed by way of parallel lines, simultaneous phrases, a spatial dispersion and regrouping, at the same time, in the same moment: "speech" is freed from breath. The line instead becomes "an assemblage."[64] Within such assemblages, lines unfold kinesthetically, defining "movement paths" that crisscross from center to margin, left to right, up and down, across the body and across the picturing of Sign—taking shape as "transformation lines," a poetic technique within ASL poetry where one picture transforms into another without a break, creating a progression of nuanced overlaps or superimpositions.[65] Lines are always "in relation" to others, and in the work of ASL poetry, become the basis for visual-spatial-kinesthetic compositions, a threading work "where lines, handshapes, movement paths, location, palm orientation, and all the other linguistic ingredients are not ends in themselves but, in their various ways, generators of the *moving image*."[66]

The assemblage of lines proceeds as a form of montage, figuring close-ups, cuts, juxtapositions, that constitute a type of "refraction"; a spatialization, a picturing, a poetics of movement paths that cinematically figure a range of viewpoints. Following Tim Ingold's reflections on the nature of lines, transformation lines may be understood to create a dynamic sense of flow by bridging distinct phrases or pictures. Stretching or pulling the line gives a sense of continuity that is also a type of "flourishing"—leading into a continuous becoming of meaning, a form of "anthropogenesis," where poetry is just as much a picturing as it is an articulation of bodily vitality: the flourishing of meaning by way of transformation lines is equally an unfolding of the body as a craft.

Tim Ingold, in reflecting upon a range of lines, provides a suggestive distinction between "doing" and "undergoing," as well as between the "labyrinth" and the "maze." These distinctions allow for probing the notion of "anthropogenesis," which captures "the life of lines" as they lead us into particular experiences. By way of anthropogenesis, Ingold aims at unfolding the line as an experience and vitality of life lived, as not only the organization of time upon an axis of past and future, but equally as what happens along the way, for "where there is human life there is never anything but happening."[67] Life, following Ingold, is always "life-in-the-making" and as such, lines become the basis for a range of makings or undergoings. Whereas one may dedicate oneself to various tasks of "doing," often setting one's sights on the accomplishment of things, life's creativity is equally "without beginning or end"; rather, "human beings both grow and are grown, undergoing histories of development and maturation."[68] In this sense, we may come to appreciate "doing" as also what one "undergoes" in the crafting and making of things, reminding of the ways in which things also shape the body, the mind, and our social relations. As Ingold asks: "Are we not always with things before we do anything to them?"[69]

The process of such meaningful experiences lends to the "anthropogenic" paradigm Ingold argues for, which is less about "mastery" and more about "humanifying" our place in the world, turning the line away from the "maze" of getting things done and toward the "labyrinth" of creative flow, of "doing-in-undergoing" that supports attuning one's perception to the signs around. While one may follow a path in the labyrinth, that path is less about a "walker's intention" (as in the maze), and more about paying "attention," and staying close to a multiplicity of views.[70]

Moving by way of the walker in the labyrinth, I'm curious to consider the gestures and expressions found in ASL poetry as an arena where lines also take on life; where one stays close to the "signs" found along the way, asking of one to pay attention within the creative process of "doing-in-undergoing." Jim Cohn, in his insightful studies in ASL poetics, highlights the differences between writing, or text-based poetics, and signed poetry. For Cohn, writing is always an abstraction, a work of concepts that may refer to entities in the world yet do so by way of abstract forms, or static lines, on a page. In contrast, ASL poetics is always situated, and as such "remain close to the living human life world."[71] Such a difference, for Cohn, acts to "liberate" language and one's relation to it, turning the "line of poetry" into a "vivid rhythmic form."[72]

Within ASL poetry, the poetic line is pulled into the air, stretched across a D/deaf space that is not so much nonlinear as it is "anthropogenic"; by putting the line into perennial motion, a stitching of pictures whose thread is continuously extended, looped over and then back, from picture to picture, ASL poetry figures an assemblage of lines that is precisely a space of living. As such, it keeps us in the labyrinth, where we walk, accompanying the movements of the poem, bringing attention to the transformative power of lines.

Haptic Life, Holding Touch

From peripheral visions and gestural voice to the tactility of bodily hearing and the lines of expressive poetics, the experiences and articulations of Deafhood open up a range of theoretical and cultural perspectives that move understandings of sound and listening, voice and language. To follow along this diagrammatic map, this labyrinth, I continue by way of a reading of Erin Manning's *Politics of Touch* (as "touch" comes forward, is dramatically at play within Deafhood: the touch of sound, of language, the touching that is movement, that is also a grip of the political).

Manning outlines touch as that which manifests a deeply processual relationality in which any form of "pre-constituted" subjectivity (Being) is overturned, placed instead within a framework of encounter, interaction, choreography: a dance of engaged pressings, brushings, caresses, of prodding as well as resistance, a pushing back. Through the highly suggestive framework of touch, Manning emphasizes how one emerges as a subject, is individuated, yet never to solidify fully. Rather, one remains malleable, processual, always already *in touch*. "Touching you, I propose to you to receive, to touch. To touch is not to manipulate. I cannot force you to touch, I cannot coerce you. . . . To touch is to tender, to be tender, to reach one tenderly."[73]

Touch is posed then as a tender theory of subjectivity and relationality, where touch is underscored as a worlding gesture giving way precisely to a ceaseless becoming, where touching "takes place as a trace, a detour, an erring."[74] To touch is to open the limits of the body, as a bordered self. Touch, therefore, is a politics, a question of power; the becoming that sets the scene for an encounter with the unknowable: "the moment when I recognize that to touch an other is to touch the impenetrable."[75]

Within D/deaf experience, touch is profoundly at work: from the tactile sensation of feeling sound ("touch-hearing"), the vibrational stirrings and movements of things that race across the skin, pressing the body, slightly or deeply, to the ways in which the hands touch language, and themselves, in the act of signing—the hands that give language a material form, a spatial expression, moving over and into words, expressions, across the body and onto the eyes. Touch is a dynamic sense, a gestural category, a perceptual, social, and linguistic capacity (leading to the notion of the "tactile mind."[76])

What becomes of a politics of touch within a framework of D/deaf experience and by way of Deafhood? In what sense does "touch as erring" occur as one signs? While Manning emphasizes touch as a gesture that "opens us to a story we have not yet heard, to an unworked work, a narrative without beginning and an end,"[77] Deafhood, in contrast, moves us toward a story that we firmly see; that, I would suggest, unworks touch precisely as what the hearing world may evoke in order to upset the capture of subjectivity. Rather, Deafhood might be seen to locate touch precisely as what grounds subjectivity within a culture—touch as what resists being stripped of language. Touch, in other words, might be what allows for defining oneself, and the stories one carries (or have been forced to carry), enabling a "telling" that may then articulate touch within a politics aimed at countering the constant figuring of oneself (by the hearing world) as having *erred*. ("I call you out for refusing to acknowledge / sign language in classrooms, for assessing / deaf students on what they can't say / instead of what they can, we did not ask to be a part / of the hearing world[78]")

Whereas Manning poses a politics of touch as that which productively "throws us off balance,"[79] a D/deaf politics of touch may be what works at equilibrium and knowability: a touching that signifies, that narrates, that figures one within communicability, a meaning (a wording, a culture) carried between subjects. A D/deaf politics of touch interrupts audism as it positions touch as what may detour subjectivity, as what may throw one off balance, allowing for alternative relations to the body. In contrast, touch articulated by D/deaf experience and culture becomes a gesture of balancing: a holding onto, giving way to something both figurative and concrete. "While the formal properties, the deep structure, of Sign allow the most abstract concepts and propositions to be expressed, its iconic or mimetic aspect allows it to be extraordinarily concrete and evocative, in a way, perhaps, which no speech can be."[80]

Haptic Narrative: Firmness

I'm led to think of firmness, and the quality of continuity: touch is what contributes to trust, reliability, restraint. Returning to Manning (in which touch is underscored by way of the desire for movement, envisioned as a constant becoming, a politics of touch as an insistence on destabilization): What becomes of becoming by way of D/deaf experience and culture? By way of the D/deaf diagram I'm sketching? Politics of touch is posed as a question of bodies in movement, bodies as movement, of departures, and yet, there are other stories touch gives. Touch is also about arrivals (ground, a terrain, a touching down);

touch as the establishing of fixed points from which to move. In this sense, identity is not so much a continual movement (as Manning insists), rather, it is an oscillation; a *gravitation* around certain points, places, relations; the making of worlds (the worlding taking place) that Manning underscores by way of touch is also about figuring a constellation of stops, rests; identity as an itinerary consisting of departures as well as arrivals: a dance that is more tension than release; that pauses even as it flies.

Identity might be something that is captured, that is housed within a system (of references, institutions, communities: "the nation" as Manning highlights); yet, it is also a process of inhabiting such fixed points: of making a home, of fighting for home. Identity is something one must rely upon in the fostering of relationships: I know you as an identity, and I trust you as an identity. I may also *take on* identities given to me (as a son, as a husband, a teacher), but I also move within and through them—I adopt them just as they adopt me; I grow into these identities, and they may also allow me to grow, as they touch me and as I touch back. As such, I am also changing along the way; my identity changes as I inhabit it: I may also grow out of particular identities, as I search for the next step, or am taken along, giving way to further points—rest stops—from which to extend myself.

And further, I also rely upon the identities of others to feel secure in the relationships that touch me and that deepen as we move together, as we go along. The processual body is therefore not a completely splayed body, but a vibration hovering around a center (a heart, a set of concerns, a name); a series of oscillations that leave the body, pulling it forward while also returning. The body, as such, is not necessarily impenetrable (as Manning poses); rather, it *gives way* to my touch, as I to it, and as such, the body is no longer figured as an impenetrable or inaccessible depth, but rather, a series of surfaces; it is not something to *get into*, but rather, by way of touch, it shows relations as a process of surfacing (rather than penetration). It may resist, in forms of self-preservation, but still, the body recedes under my touch; it is soft as it is hard—it flexes, this body, to press back upon me even as I resist: even as I question. This is precisely what I carry, in the memory of someone: the degree to which touch has given way to something else, to something less physical, less bodily in fact; a knowing, over time; intimacy, emerging over the time of the movement we are, in the taking and giving one experiences, together; and in the folding and unfolding of surfaces. (This is how we become dance partners: knowing the way another moves, will move, anticipating each other by way of a history of movement.)

Potentiality—this becoming—then sits within a framework of actuality: while potentiality is always already present, in the touching we perform, and that defines us, it nonetheless resides within a set of ordinary rhythms that give pace to such potentiality; that defines, through its weak or strict structures, such as the state, the institution, or language, the processes of emergence—the potential (and politics) of touch is precisely how it gains movement from the actual, from within the coordinates of a given framework.

Touching, being touched, being involved in touch . . .

And, then, I also touch myself: while touching another (you, this thing, matter, sound) may rupture and fulfill me precisely by way of relationality, reminding of my contact with the world as a defining performance, an intensity (shattered, interrupted, loved), I also return to myself, in the touching I give to myself as myself; while I am brought into shape by the touch of others, in the passing of contacts, in the lessons of the caress, the bruise, the aging, I also come back to myself, by touching myself: I speak to myself, I wrap myself in myself, I resist the world and the touch of others; I hover around myself, even while others take of me for themselves.

The politics of touch is precisely the "haptic articulations" that enable both becoming *and* being, both flight *and* rest, a movement away that may also return us, precisely to the firmness of relation: this moment when, during the night, I reach over knowing your hand will be there, knowing the feeling of your hand precisely as what I will feel again, what I search for, in this moment of needing your touch: knowing what your touch is precisely because I have touched it before, as it responds under my hand, again and again.

I know by way of touch because touch also stays the same.

* * *

Reading and re-reading Erin Manning's *Politics of Touch*, I'm led to consider how D/deaf touch is profoundly firm, linguistically articulated, leading toward a multimodal sensory world that is poignantly narrated as a drive toward identity, toward a recognized cultural identity working at community. As such, D/deaf touch contrasts sharply with Manning's approach. Consequently, I would say that I learn from both these forms of touch; from the erring potentiality Manning works at in order to reorient a sense for language and nation, for being a citizen, keeping one in touch with others, to the cultural experiences and visions articulated by Deafhood, that lives a politics of touch precisely as a fight for the right to live in the world, as a stable ground—what Jacobus tenBroek would argue for, challenging legal assumptions around disability.[81] I learn from both these forms of touch, and accordingly, I'm led to speculate as to their meeting and overlap—which may give way to a "listening through the bone" that equally enables new "bilingual" or "bihaptical" constructs: a citizenry that works at those interdependencies and tender worlds Manning suggests and that may also reorient the world of the hearing by grounding theory within the haptic narratives of signed subjectivity.

Vibrational Worlding (Pedagogy)

I want to approach the question of haptic life, and following what Jim Cohn terms "sign mind,"[82] as an articulation of touch, a politics of touch, and also as a vibrational worlding, which D/deaf experience and D/deaf subjectivity seems to articulate: a haptic form of mind, which also emerges in ever new ways within contemporary life, from social media to the smart device. A vibrational worlding figured in the state of connectivity, giving way to a cognitive materiality.

> A set of (vibrational) notes: . . . haptic listening, haptic media / online, texting, etc. / voice in the device; in artificial bodies: question of an acoustics of network culture; acoustics of digital experience and processes (fake news, online dating, network identities) / voice recognition politics (Siri, Alexa, Google) / touchscreen, touchpad: "cognitive capitalism" under the hand / typing, texting, touching always already linked to the algorithm / what becomes of "mind" (by way of vibrational worlding: the "ecology of feeling" and the economy of affect) / tap . . . tap . . . / voice in the hand, the listening hand / a new "grapheme": grammatology of the finger (the finger-print: another form of publishing, a being-public: press, tap, press, tap: access) / these sensations that pass across the tiny grooves of the hand, as the network responds (Tiziana Terranova: "the network as affective vibration"[83]) / Aaron Williamson, in his performances, undercuts the ear as an "aspirational" organ, as the heart of poetry; rather, what of the breath? (recalling Gil J. Wolman's breath-poems—before language, after language: breath as poetic vehicle) / Williamson gestures, and leads us: what of the wind in the lungs, the beating heart, the blood behind the ear? (John Cage's anechoic experience: "hearing" the rush of blood and nerves within the body: a silence that may give way to Deaf music?) to the thrush and rush, flutter and

> shudder, as a respiratory refrain: this vibrational humming of the body—the vocal folds threaded through the handshape handspun handbreath: sculpting air / and then the knees, the feet, which Williamson performs, follows, writes with: the feet across the floor like a gauze giving way (ripple bend flex);[84] the throat of the feet / it travels upward, along the tendons, the calf and behind the knee, upward along the thigh tensed with stillness, a language of the muscle, to bloom with the sudden about face and then down, back twisted shoulders raised / "to be at the edge of language" / at the edge of hearing / this is where Williamson starts, listens, moves: vibrating poem.

[break]

José Gil, in his book *The Metamorphoses of the Body*, gives a description of the body by way of "exfoliation": the spatial transfigurations inaugurated by a body in movement, where the body "spatializes space,"[85] gives it articulation. A type of "shedding" continually at play and by which spatio-corporeal limits are stated and then modulated, breached, reworked, for the body is always already "in touch" with things, scraping against, brushing past, being pressed upon. For Gil, the body is always already a question of space, and of movement: a bodily schema/figure-ground. The body is therefore articulated by way of "levels of language," a grammar that fluctuates, moving from articulation to the "below of language," where the body speaks—an "infralanguage" of sensation, of transductions, of passages and of gesture, what Gil terms a "preverbal meaning matrix."[86] From within an infralinguistic framework, the body emerges, is emergent, articulated, leading Gil toward notions of "dance" as that arena where exfoliations, and the rhythms of transfiguration, are drawn out, showing the "intelligence of the body."[87]

[and then, back, connect]

> A breath/a respiration/which is the constant extension, and then return, that is always the crossing, this breath/Where are the limits of the body then? In this constant dance of respiration that draws the air into oneself (which is always unimaginable: I can never comprehend the air that is already in my lungs), and then pushes it back out (to where, for whom?): I must pause here, on this border that is only constituted by being breached, by being moved—the metamorphosis of the body as it respires, as it lives, figuring a body-space that brings touch into a framework of breath (a biopower): the touch of the wind upon the ear, the touch of breathing as my chest rises, then falls/such is

the "intelligence of the body"—this dance, this infralanguage, this vibrational worlding that is breath always already at work: a sensing oneself as always already respired, a movement that is steady rhythm, an infrarhythm, that is more *figureground*/as Aaron Williamson opens his mouth to show his breath in place of voice: to turn the deaf ear into a lung, a breathing by way of the cochlea.

His movements that draw from the infralinguistic, drawing it through infrarhythms: the Deaf vibrational poetics whose dance is radically exfoliant: inhabiting and articulating the frequencies of haptic life.

Deaf Science: The Haptic Mind

What might emerge here, within this vibrational worlding, by way of infralanguage, as what may give expression to a synesthesia of the senses, a poetics of plasticity, and the cross-modal currents always passing within the brain, across the ear, along the hand? What type of mindfulness may be harnessed, or practiced, by attending to vibrational worlding, the sightlines of Deaf space, the peripheral attunements and dynamics of touching as both flight and rest? To listen by way of the bones—a listening that becomes suggestive for considering the "haptic mind" of D/deaf subjectivity and textured by the infralinguistic. And which seems to find correspondence with the mind emerging by way of haptic media and the liquid, vibrational economies traversing the globe. The biopolitical foundation of contemporary capitalism, as that which figures life itself as a form of labor, as currency—the endless event which, as Franco "Bifo" Berardi poses, puts the soul to work.[88] Such "biocapitalism" finds its operative footing through an intensification of precarity, which exhausts the body and undermines conditions of (public) support, as well as through a form of cognitive capture, suggesting a new class which Berardi terms "the cognitariat." Cognitive capitalism, as an instrumentalization of biopower, figures "the mind" within a systemic flow (or bombardment) of energy and information—an affective capture manifesting in a cognitariat wedded to and stressed by the network. Hence, we may appreciate the intensification of concerns for "mindfulness" and mindful learning today.[89] From meditation practices to new articulations of somatic care which circulate increasingly through the cultures of the cognitariat, all point to the ways in which the haptic mind relates itself to contemporary life and politics: that in the biocapitalistic matrix, which houses the body and mind in a paradigm of work, laboring at the economy of events or the insecure flow of freelance life, new ways of processing and balancing are needed. To specifically attune to the body's limits, as well as to common the resource of life in general (to establish networks of care), so as to figure more sustainable approaches. The haptic mind, as that mind shaped by experiences of touch, the vibrational worlding, and infralinguistic exfoliations effecting what it means to be in this world—moving in the world, brushing against, hitting, and feeling precisely the edges as well as openings, this state of connectivity, which is also easier for some more than others (the network is always uneven)—such tactility may come to act as a common foundation for keeping one in touch with the bones as well as

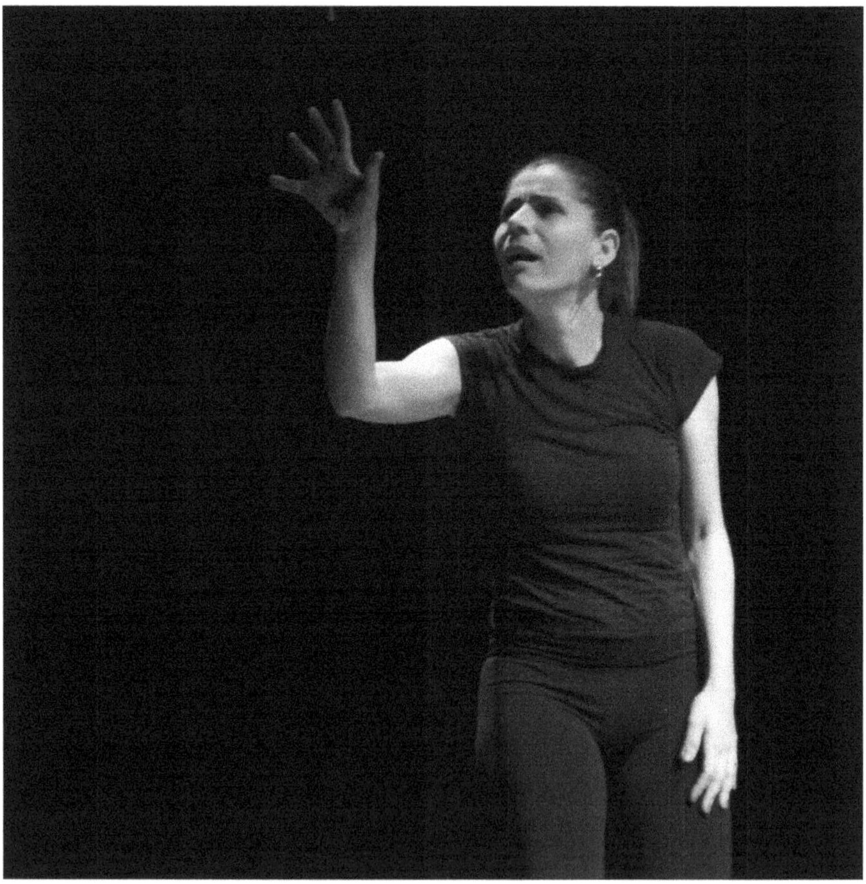

Figure 4.3 Mariana Santos, in "Surdo, Logo Existo!" ("Deaf, therefore I exist!"), Teatro Guaíra, Curitiba, September 2019. Photo: Gilson Camargo. The sign she performs is of a broken dream, a lost hope, *the impossibility of communication.*

the soul. A common, considered touch finding guidance from the haptic mind of Deaf perception and subjectivity. A Deaf science that may provide (synesthetic, vibrational) tools, or (infra)linguistic grammars of haptic life, so as to negotiate the politics of biocapitalism and its cognitive capture.

Such perspectives equally speak toward a bioacoustic framework, and the ways in which medical approaches toward deafness, and disability in general, often adopt methods of technological enhancement, such as cochlear implants, impacting onto conceptualizations of *life by way of hearing*. This extends to questions of bioengineering, and the capacity to genetically modify the human body. Medical approaches to possibly eliminating deafness thus draw out profound ethical debates. From biocapitalism, and the affective networks conducting much of global life, to bioacoustics and issues around engineering "hearing persons," such realities come to articulate a larger ground onto which a Deaf science offers important input.

Acoustic Justice and Deafhood: Plasticity, Diversity

I have been considering deafness and Deaf culture as a framework that can enable a number of critical pathways within Sound studies, specifically with a view toward elaborating a theory of acoustic justice. Given that histories of D/deafness are marked by a range of deep challenges and struggles, whether in confronting forms of oppression, stigmatization, and ostracization to more radical interventions often driven by medical and educational ideologies, including eugenic campaigns in the late nineteenth century and the forceful mainstreaming of the deaf through oral education, it becomes clear that D/deafness may teach us a great deal about the ways in which the meanings of sound, listening, voice, and language wield great social, political, moral, and cultural force.[90]

In his book *Understanding Deaf Culture: In Search of Deafhood*, Paddy Ladd looks at histories of D/deafness not only with the aim of chronicling the particular struggles that have shaped Deaf culture but also by paying attention to the stories and experiences of D/deaf individuals and communities, which often lead to appreciating the smaller resistances emerging from within everyday challenges. As Ladd seeks to highlight through the term "Deafhood," cultures of the D/deaf are rich with knowledge about the ideologies and habits shaping human communication and the presumed qualities or capabilities that all voices and bodies should demonstrate. Deafhood, rather, is built upon creative and critical negotiations with the legacies of oralist and audist assumptions and policies (educational, medical, social) and which come to provide important guides for understanding Deaf culture as well as hearing culture. Ladd points the way for working at a decolonization that takes aim at audism and oralism as colonial forces (against the Deaf nation) and which can undercut dominant notions of language as being bound to verbal speech. The decolonizing project of Deafhood not only supports other forms of voice, and other understandings of hearing and listening, it further wields a critical consideration of currents in medical practice that continue to see deafness in terms of disability and deficiency.

CIs, and other medical research aimed at "restoring" hearing, are certainly one of the more contentious issues within Deaf culture and studies. The Alexander Graham Bell Association for the Deaf and Hard of Hearing, a long-standing player in debates around deafness, focuses exclusively on promoting "oralism" (speaking over signing), and educating people, often hearing parents of deaf children, about the benefits of CIs. Drawing references to the long history of struggles over the "right to sign," including the significant Milan Conference of 1880, which saw the eradication of more progressive approaches to D/deaf education and culture, it is understandable how CIs are seen to perpetuate genealogies of prejudice and violence facing the D/deaf. As Ladd forcibly suggests, "a public climate has been manufactured [by oralists] which once again proclaims the advent of the 'miracle cure' and suggests once more that 'deafness is abolished.'"[91] CIs raise important questions as to the future of the Deaf community. For research has shown that as implantation technology advances it will continue to impact onto how D/deafness is approached by the "oralist" or medical establishment, and for the D/deaf as well.[92] Deaf culture, in this sense, must continue to navigate the evolving medical, political, and technological terrain.

Central to cochlear implantation is the issue of language acquisition, and the capacity for deaf children to access verbal speech (as well as the dominant hearing world). Having the capacity to hear, particularly as an infant, becomes central to increasing chances for acquiring spoken language, and therefore, to increase an individual's opportunities for participating in the hearing world and its institutions. Within recent proposals and policies, some Deaf organizations and institutes, such as The Learning Center for the Deaf, seek ways of accommodating the new reality of implants, understanding that the practice of implantation has defined a new D/deaf subjectivity and culture.[93] The implanted person, now, signals a new D/deaf experience, for even though one may use an implant, one is still physically deaf. Sounds need processing; as they are picked up externally and then processed and transmitted through a set of implanted electrodes to the auditory nerve, an individual must learn how to "hear" and comprehend the limited signal information.[94] In adjusting to this new form of hearing, a CI user must also navigate their new sense of identity situated between D/deafness and hearing. This impacts in many ways on physical and cultural understandings of what it means to be D/deaf, as well as on bioacoustic approaches (in terms of conceptualizing hearing life). As Mara Mills outlines, new types of intersections start to emerge, suggesting other perspectives as to what counts as hearing and non-hearing, and what may constitute Deaf life in the future.[95]

While it is imperative to remain engaged in the ongoing struggles and issues in terms of arguing for human diversity, it also seems important to consider how a new manifestation of D/deaf subjectivity captures possibilities for bilingualism, and bicultural lives, by crossing spoken and signed languages and worlds. "Deaf children with CIs who are exposed to both visual and auditory languages can be described as bimodal bilinguals—individuals who use two different languages in two different modes."[96] Bimodal bilinguals include hearing individuals who use spoken and signed languages as well as D/deaf individuals who communicate using signed languages, and also who may be able to speak by accessing sound through a CI.

The presence of bimodal bilinguals, and subsequent new expressions of Deaf identity, presents opportunities and important questions. As Jane Dillehay argues, CIs carry significant ethical challenges, especially when it comes to conducting surgical interventions on otherwise healthy infants and children.[97] Such acts beg for a deeper tussle with the prevailing ideologies that understand such "modifications" as essential to leading a "normal" life. In addition, CIs intersect with debates around genetic research, and medical approaches to genetic intervention. The possibility of "removing" deafness through genetic screening poses important challenges to the D/deaf community as well as to essential ethical and political debates.[98] Questions around bioethics and bioengineering further intersect with arguments about neurodiversity, which underscores deafness as part of the range of human diversity. In addition, as Mills points out, CI usage also brings into the D/deaf experience and community a "politics of the signal," highlighting how advancements in technology surrounding implants (and their signal processors) are equally an issue of sensory politics.[99]

How to navigate through the ethical, political, and medical challenges? Are there ways in which the new expressions and identities within Deaf culture can give crucial input in approaching technological and medical cultures and their sensory and bioacoustic politics? In what ways might neurodiversity arguments shape bioethical questions within D/deaf communities? Are there ways to think about CIs as participating within rather than undermining neurodiversity debates by explicitly manifesting a new type of identity and body, one figured between hearing and deafness? Might Deaf persons take a lead on debates around "brain plasticity" or "the mindful brain" by drawing from the knowledges and insights of the D/deaf community, which has always had to combat presumptions around what constitutes intelligence and ability? As well as offering insight into establishing more mindful approaches to what it means to live in a world defined by signals?

In this sense, I'm curious in what ways the concept of listening may add to the discussion. Returning to Gemma Corradi Fiumara, and her understanding of "the responsibility of listening," I'm led to consider how listening is dynamically multimodal, moving across our physical *and* ethical senses, and which gives way to paths of knowledge by bridging biology and being, the inner voice of bodily intuition, as well as the capacity to "listen through the bones," with the capacities of thought, language, and social understanding; a knowledge that also integrates an understanding for how technologies have always challenged and elaborated listening's reach. In this way, an understanding of the multimodal and interpersonal capacities of listening, its creativity and plasticity, might participate in supporting collaborations across diverse subject positions.

Understandings of listening seem to already suggest a highly expanded framework that specifically articulates sensitivity for making connections between ourselves and others, between sensual life and ways of making sense—which Anna Tsing underscores as equally helpful in shaping a more justly connected planet: to find an approach to recognizing the "polyphony" of earthly systems.[100] In this regard, listening is understood as being untethered from the ear, as well as for its capacity to care for a greater worldview. And from a D/deaf perspective, is greatly articulated outside the framework of aurality. By critically highlighting listening beyond the strictly aural, the hearing world may continue to learn from the D/deaf who, in seeing meaning everywhere, can also give lessons in looking out for each other. The "acoustic separations" that Péter Molnár works against, in the context of Roma communities in Hungary, can be extended here, to suggest a more pronounced and generative approach to listening as what may help in bridging the divides between the hearing and the D/deaf, and by extension, the implanted and the non-implanted.

As I've been arguing, D/deafness can be emphasized as what *allows* a broadening of dominant understandings of sound and hearing by explicitly refocusing hearing (and voice) around a range of suggestive coordinates: from the visual-gestural languages of Sign, and the visual and spatial intelligence that establishes by way of D/deaf space extremely nurturing and dynamic views, to the unique poetics articulated by way of ASL poetry which gives so much life to the meanings of gesture and embodied expression, and that can be seen to further capture a politics of D/deaf touch and its ability to grasp other forms of knowing; and from which a sense for the "haptic mind" of D/deaf cognition may suggest extremely relevant approaches for contending and engaging with our contemporary cognitive economies and globalized network life: the "affective vibrations" underpinning our relationships, labors, transactions, and their mediated currents. In this sense, it becomes important to also recognize the degree to which D/deaf individuals and communities have always navigated between hearing and non-hearing, speaking and signing, figuring techniques and methods for living (and surviving) between different worlds. (This becomes clear in terms of recognizing how Deaf families are often mixed, including both hearing and Deaf members.) From "signing undercover" or developing "home signs," to lip or speech reading, from seeing sound and touching words, recreating architectures or building (and hacking) assistive technologies, or establishing important cultural and social institutions, such as Deaf clubs and Deaf theaters, not to mention continually mobilizing for equal access and disability rights, Deafhood is founded on defining creative and critical routes in and around the hearing world. The complex realities around CI usage can only be added to this terrain.

Returning to the question of acoustic justice, D/deaf culture contributes a vital input in upsetting normative understandings of sound as well as the disabled ear. In this way, D/deafness and Deaf culture bring important critical interruptions onto Sound studies, as well as some of my own thinking around acoustics, challenging tendencies that might overemphasize "emancipatory" potentialities by way of sound and hearing. Deafness may remind such tendencies, and such views, that even while struggles over recognition may be captured through forms of articulated sound, the larger framework of auditory culture also carries the injustices of audism, along with assumptions around the "uniqueness" that only the acoustic sphere can deliver. Following D/deafness and Deaf culture allows for challenging approaches toward researching sound, as well as deepening a sense for what acoustic justice might be about, firmly reminding how listening is equally expressed through acts of looking.

Additionally, Deaf culture and struggles articulate perspectives on acoustic justice by outlining the concept of bioacoustics, which reveals the ways in which "the body" may be captured as an auditory question. As such, conceptualizations of *life by way of hearing* greatly underpin oralist or audist ideologies, figuring a bioacoustic politics. Technologies and medical approaches toward assistance and enhancement, from hearing aids to CIs, lead further to questions of genetic engineering, and as such, extend beyond the individual body to impact more forcefully on all forms of life. From an acoustic justice perspective, how we understand sound and hearing thus leads to larger discussions and concerns, from understandings of ability as well as life. I follow Jonathan Sterne in this regard, and his suggestion that Sound studies should find a way of engaging more fully with Deaf studies as a way to complicate and elaborate discourses on sound and listening, as well as for thinking through "the plasticity of the human subject."[101]

Christine Sun Kim, a Deaf artist working in the context of sound art, underscores sound as a form of social currency: a phonocentric economy by which one gains fuller access to institutions and the greater social arena.[102] Through a range of artistic expressions, from drawing to performance to public projects, Kim works at challenging the flows of such a currency; redirecting its ideological powers, and its institutionalized rhythms, sound is captured and articulated from a Deaf position and perspective. In doing so, Kim articulates a critique by way of D/deafness, deconstructing or destabilizing the field of sound art while also revealing how sound is equally expressed by way of movement or visuality. As a consequence, Kim makes clear the ways in which hearing carries value, and as such, comes to limit an understanding of ability. In this regard, her work forms an important act in chipping away at the acoustic separations placed between hearing and Deaf worlds.

I'm curious to consider in what ways such a critique starts to figure the social currency of sound differently, and the power it may wield. Following this creative and critical intervention, this performative upset, might the currency of sound be seen to take on other functions, as a currency in support of forms of bilingualism or bicultural movement? Kim suggests as much, often drawing parallels between ASL and musical performance, for instance, where the dynamics of hand movement are fundamental to the articulations of meaning.[103]

Yet, by articulating sound, and sound art, from a Deaf perspective, we may also begin to detect other routes, other cultural identities and experiences that, in fact, point beyond hearing or not-hearing as stable positions. In other words, what might be the fuller consequences of such crossover, in terms of subjectivity, cultural identity, and language?

The curatorial project, *Infinite Ear*, developed by the art organization Council, focuses on similar questions and issues. In particular, the project seeks to foster exchanges between hearing and Deaf worlds, and offers an expression of the crossovers I'm considering.[104] This is based upon, among others, an investigation into music, and the musical instrument, as an arena that too often excludes the D/deaf, or that gravitates around specific phono-normative understandings and assumptions, for instance that sound is only accessible through the ear. Importantly, for Council, the project aims at upsetting such perspectives onto listening and musical experience by researching and developing musical instruments (led by the artist Tarek Atoui) specifically designed in collaboration with D/deaf persons, as well as hearing professionals and designers. This results in the production of a range of experimental instruments and sonic experiences, often focusing on vibration and visualizations of sound as critical links between hearing and D/deaf persons.

Such interactions and cross-cultural expressions may also speak toward the greater history of Deaf theater work, which has often been about bringing together hearing and Deaf performers and audiences. From the establishment of the National Theater of the Deaf in 1967, in the United States, Deaf theater has flourished into an international community. Deaf theater has charted an important path in nurturing Deaf culture, as well as challenging traditional understandings of theater as being a space of orality and audibility, a scene defined by the staging of voice. For example, the recent theater production by Grupo Mapas, "Deaf, therefore I exist!" staged at Teatro Guaíra, in Curitiba, Brazil, speaks toward histories of Deaf subjugation, while also enacting a form of emancipatory takeover of the stage: a celebratory, Deaf occupation of the theater that interrupts the stage as a site of audibility.[105]

Following these cultural expressions, from Christine Sun Kim and the project by Council to "Deaf, therefore I exist!", the currencies of sound and hearing begin to emerge as critical contact zones, as bicultural and bioacoustic flows that create surprising intersections. Such cultural collaborations come to articulate a sensory politics as well, impacting onto established constructs of the visual, performative, and musical arts. In this regard, bringing a critical and creative view onto "audio supremacy" allows for thinking through or around the distinctions we might make between hearing and D/deaf worlds and subjectivities, not so much to arrive at a third term (or third ear) that might resolve the conflict, and the real ethical and social challenges around CIs for example, but rather to engage with the "manners of living" that articulate themselves by way of diverse modes of hearing, communicating, and sharing. As Tiqqun states in the opening lines of their *Introduction to Civil War*: "The elementary human unity is not the *body*—the individual—but the form-of-life."[106]

Deaf culture can significantly contribute to a sensory politics, a (post)colonial politics, a technological bioacoustic politics, as well as a cultural world of diversity, by elaborating forms and methods of sensing and speaking, learning and living. To wield a particular knowledge in the making of more mindful institutions or culturally dynamic environments. Deaf culture, in this way, can greatly impact onto what we may understand through acoustic justice, pointing toward particular perceptual and cognitive capacities, and different ways of grasping and building the world—different ways of intervening within the distribution of the heard. As such, it also significantly elaborates what it means to "have a voice," in the struggles over recognition and social equality, explicitly moving across individual bodies, and their abilities, to forms of life—the manners of living defined less by "hearing loss" and more by "deaf gain."[107] What is gained, as I've tried to articulate, is a range of perspectives and approaches to perceiving and articulating relationships, to wielding a politics and poetics of speech and hearing, seeing, and touching, which may continue to contribute to what we understand these to be. And importantly, for the project of acoustic justice, diversifying not only how we understand hearing but also our manners of living.

Acoustic Support

I have been pursuing questions of relationality and recognition, attention and sensuality, community and justice, by way of a focus on acoustics. Acoustics is brought forward as a generative base from which to approach how people work at capturing a sense of social possibility. This leads to considering the greater conditions of which acoustics is a part, which lend to shaping relationships and our capacities as individuals—from the physical and material conditions of architecture and the built environment to the social and political frameworks that situate one within particular histories and discourses, emotional and psychological experiences. In this sense, acoustics is elaborated as a means for contending with one's situatedness, functioning as a performative mechanism by which to configure or reconfigure one's sense of placement or belonging. Through a range of acoustic gestures and practices, one may work through how it is that one is positioned within a given social or cultural arena. This entails appreciating acoustics, and the related articulations and movements of sound and listening, as fundamentally embedded within our capacity for orientation: for balancing and for synchronizing, connecting and conjoining, for working around alignments and misalignments.

Following this approach to acoustics, acoustic justice is offered as a framework for addressing this complexity, underscoring how sound and listening impact onto experiencing or articulating a sense of agency: for capturing possibilities for attunement or interruption, for defining pathways of transformation and togetherness. Specifically, acoustic justice is envisioned as being able to give traction to struggles over recognition, while also challenging how we understand social equality. Acoustic justice, and the ways in which sound and listening extend and entangle human and nonhuman, politics and matter, equally considers greater ecological and posthuman perspectives that necessarily exceed or transform what constitutes sociality. This emerges as a more general concern within my project: to consider the ways in which an acoustic model allows for a broader, more *thick* approach to understanding subjectivity, articulations of agency, and the interweave of sensuality and relations of justice. Through an acoustic model, a pronounced understanding for the dizzying intersections and interdependencies defining our material and social worlds becomes possible. In this sense, sound and listening, hearing and voicing, and how they participate in facilitating a deeply expanded perspective on relational bonds, emerge as resources.

I've worked through mapping and detailing this acoustic model throughout the preceding chapters. Moving from issues of physical space, and how acoustics is fundamentally tied to questions and experiences of place, I put into play the "reverberant geometries" of sonic experience and aural architecture. This allows for understanding the particular qualities and potentialities of acoustic practices and performativity, and how formations and configurations of common worlds, their self-organized compositions and improvised acts, capture modes of agency and collaboration through an acoustic politics: the interrupting of dominant tonalities and partitions of the heard in support of participatory governance. At the same time, reverberant geometries and acoustic practices are readily applied to more harmful and violent scenes and projects, revealing how acoustics may be weaponized. From such a view, acoustic justice provides a greater understanding for the ways in which the built environment, social well-being, and a politics of recognition and rights become acoustic questions.

Extending from a focus on space and architecture, from acoustic commons to acoustic violence, I work at elaborating the particular relational qualities and affordances that are brought forward through acoustics. In what sense do the reverberant geometries or transmitted interferences particular to acoustic space exceed the limits of architecture to facilitate and incite a range of connections, from human to nonhuman, from the bioacoustic to the electroacoustic? A logic of transformation is posed as fundamental to such acoustic connectivity and propensity, one that I articulate by reflecting upon the field of sound art. Sound art can be heard to manifest a deeply experimental relationality, spiriting contact with animal species and the worlds of matter and energy, including the crafting of particular methods, which draw out possibilities for more affective and "resonant" experiences and understanding. Poetic ecologies of resonance emerge, specifically mobilizing encounters with the unrecognizable, the strange, the untold, or the held behind. These encounters, articulated by way of sonic compositions, field recording projects, sited improvisations, as well as through performances of singing and screaming, force a poetics into the arena of audibility. Poetics and resonance draw listening into a transformative, even transgressive state, capturing acoustics as the basis for engendering a radical movement, one that finds dialogue with the sonic fictions and cultural mixes of the black radical tradition.[1] By way of the transformative potentiality of acoustic relations, and the poetic ecologies whose resonating powers force all types of migrations and intersections, to spark strange alliances and extra-relational compositions, listening moves toward that of "dreaming." A listening-dreaming is posed as a form of engagement—a power of attention—that integrates a more dynamic sense for the emergent and often restless imaginaries underpinning social and political life. As such, understandings of recognition are reworked, put on the move, allowing for a more diffracted or resonant perspective on the relational.

From a position of acoustic relationality, a politics of location and identity plays out, forcing an understanding of sound and listening as fundamentally a question of touch: how the body is oriented and emplaced by way of existing lines of discourse, power, and sociality. Such orientations come to greatly impact onto what we hear and how we are heard, what we touch and are touched by. I pursue questions of touch and orientation by way of the skin, following the theory of "the skin-ego" posed by Didier Anzieu.[2] In what sense does the "touch of sound" afford mediation between the body and its situatedness? How might acoustics incite a critical tussle with the phenomenological and social shape of the world? If sound is a force that straddles language, voice, sociality and the affective, tactile, and bodily, in what ways does it afford maneuvering through the structures that seek to emplace us, locating us by way of the skin and the articulations of identity? I follow such questioning by thinking through what I term "queer acoustics," as a "work of reorientation": how, in seeking other bodily figuring, in working at repositioning one's emplacement, in order to reconfigure forms of consensuality, a trans-subjective touch, one equally works at an acoustic dimension, that of intimacy, ecstasy, solidarity—forms of "common sense." A queer acoustics is articulated in order to work through skin, touch, and consensuality as key to political subjectivity. This is elaborated by reflecting upon the skin as a border, which is dramatically figured through contemporary mechanisms of border control and forms of identity profiling. By engaging with the border, and bordering practices, I articulate an acoustic justice framework as what may address how bodily affect and sensuality are sutured to the political. Finding other rhythms, attuning to particular tonalities and narratives, sharing by way of consensual constructs, and the making of vibrational ecologies of feeling, acoustics facilitates alignments and synchronizations in support of unexpected movements of community.

Questions of acoustics, from architecture and place to sound art and a politics of location and recognition, are finally posed in relation to Deaf culture. Drawing into consideration D/deaf experience, and the particular expressions, languages, and cultural practices of Deaf culture, I try and complicate understandings of sound and listening, and what "having a voice" might mean in the context of D/deafness. Through investigations into sign language, D/deaf perception, D/deaf space, and poetics, I draw out acoustic justice in order to contend with discriminations against the diversely abled. The legacies of audism and oralist ideologies, which have significantly defined D/deaf experience, bring forward a critical view onto what's at stake in how we understand sound and hearing, voice and language. In following D/deafness, I work at complicating as well as elaborating such understandings, exploring ways of spiriting collaborations and intersections across hearing and D/deaf, speaking and signing worlds. The understanding of listening as what affords contact with the nonhuman, with the world of matter, and what can assist in working at critical bodily and social reorientations, figuring a politics of location and recognition, is situated alongside Deaf studies and experience in order to raise critical thinking. From such a view, assumptions around having a voice and being heard are problematized.

In working at acoustic justice, I raise the notion of "acoustic rights," or principles, including questions of the right to free speech, and the right or principle of reply, supplementing these by underscoring listening as a fundamental freedom. The "right to listen" is offered as a path for exposing the ways in which listening is figured within struggles over recognition, while itself being a political issue. The freedom of speech, or the right to free speech and expression, brings forward the importance for not only what it means to "have a voice," but equally, what's at stake in setting the social and political conditions that enable such a voice to resound in all its diversity. From my perspective, this is what acoustic justice is about: the material, institutional, social, and affective conditions that enable the articulation of sounds and voices, orientations and reorientations, attunements and alignments, in such a way as to impact onto the normative shape of recognition. I've tried to push this understanding, outlining a range of acoustic practices and gestures, and their participation in the tensed economies of attention and desire, from the hearing to the non-hearing, the spoken and the signed, including a concern for the situated or embodied dimensions and experiences, from love, attachment, and solidarity to daily conversations, shared joys, and conflicts over place. Acoustic justice is therefore equally about reflecting upon the ways in which sound and listening wield certain authority when it comes to a politics of the senses and what we understand as the able body.

Importantly, acoustic rights, and the dynamics of listening, are never only on the side of emancipatory movements, never only supportive of commoning, and the interpersonal sharing found in daily struggles and joys. Equally, acoustic justice is situated in relation to all the forces and systems, all the structures and initiatives that work on behalf of exclusion, extraction, and enclosure, not to mention pure violence. These systems or initiatives—for instance found in Syria at the Saydnaya prison, or the detention regimes of Australia and the Global North, or those against the occupied squares and squats in Madrid and Athens, and further, within the legacies of audism and oralist practices that have regularly oppressed the Deaf on the basis of an auditory prejudice, a bioacoustic ideology—equally capture acoustics, the flows and punctuations of sound and listening, as devices: in these instances, for dividing, capturing, traumatizing, and regulating people. The potential violence of acoustic regimes is thus something to be reckoned with. It is my aim to argue for acoustic justice, and the notion of the right to listen, in order to contribute to such reckoning, as well as to support the power of people as a force that finds traction by way of acoustic matters and imaginaries.

From the silences and noises that punctuate the arena of public life, to the vibrations and rhythms through which particular alignments and synchronizations, intimacies and passions, may be sought and configured, or in the echoes and transmissions that allow for refusals, differentiations, and the projecting of all types of desires and fantasies, strange and beautiful, acoustics manifests the distribution of the heard as a scene of dissonance. To work at acoustics is to work at orientation and reorientation, from the psychoacoustic, as the experiences of hearing in all their diversity, to the social acoustic and the shaping of community life, which integrates a sense for the electroacoustic, and the diffusion of mediated sound, from the vibrational to the ultrasonic, as well as the bioacoustic, the entanglements of human and nonhuman life that gives way to conceptualizations of *life by way of hearing*. Such an acoustic lexicon may start to suggest a ground from which to engage a range of questions, including that of architectural and spatial design as well as spatial experience; issues of language and voice, and the articulations by which agency and community are captured, practiced; the affective economies and knowledges shaping our capacities for care and consideration; and the challenges around governance, where having a voice and being heard figure not only as the basis for recognition and social justice but also for the necessity to reorient a relation to the planetary systems around us.

Acoustic justice, in this way, may equally be about "strategies of civility" which, as Étienne Balibar suggests, offer "another way of thinking politics."[3] It is my hope that the issues and perspectives I've set out to articulate in this book can contribute to manifesting such strategies. In this sense, acoustic justice may be cast as a pedagogy, not necessarily in the form of a lesson plan, or a discipline itself, but rather as the figuring of questions. To pose acoustics precisely as a question, which requires collective attention, reflection, and enactment: a considered doing. A pedagogy that may, in fact, travel through all types of disciplines and spaces, reminding one of the powers of attention and the resonant reach of an acoustic arena within any setting. A pedagogy, then, guided by acoustic care, of slow attention and nurturing sightlines, as a support for discovering and composing together.

Notes

Holding, Healing, Attending

1 Habib Battah, "An Open Mic Revolution in Lebanon," in Al Jazeera, October 25, 2019. https://www.aljazeera.com/indepth/opinion/open-mic-revolution-lebanon-191025150455130.html. Accessed March 10, 2020.

2 Timour Azhari, "Defiant Protestors Hold Rival Parade on Lebanon Independence Day," in Al Jazeera, November 23, 2019. https://www.aljazeera.com/news/2019/11/defiant-protesters-hold-rival-parade-lebanon-independence-day-191122202937296.html. Accessed March 10, 2020.

3 Habib Battah, "An Open Mic Revolution in Lebanon."

4 Jan Masschelein, "E-ducating the Gaze: The Idea of a Poor Pedagogy," in *Ethics and Education*, 5:1 (2010): 43–53.

5 Habib Battah, "A New Politics is Rising in Lebanon," in Al Jazeera, November 15, 2019. https://www.aljazeera.com/indepth/features/politics-rising-lebanon-191114141631074.html. Accessed March 10, 2020.

6 In *The Force of Listening*, Lucia Farinati and Claudia Firth map out the critical act of listening as key to forms of social solidarity and activism. See Lucia Farinati and Claudia Firth, *The Force of Listening* (Berlin: Errant Bodies Press, 2017).

7 Jan Masschelein, "E-ducating the Gaze: The Idea of a Poor Pedagogy," 44.

8 Étienne Balibar, *Violence and Civility: On the Limits of Political Philosophy* (New York: Columbia University Press, 2016).

9 Dimitris Papadopoulos, *Experimental Practice: Technoscience, Alterontologies, and More-Than-Social Movements* (Durham: Duke University Press, 2018), 202–5.

10 Ibid., 204.

11 AbdouMaliq Simone, "People as Infrastructure: Intersecting Fragments in Johannesburg," *Public Culture*, 16:3 (2004): 407–29.

12 Ibid., 410.

13 See Jacques Rancière, *The Politics of Aesthetics* (London: Bloomsbury, 2013).

14 Mojca Piškor, "Reading the Sound, Listening through Words: Resonances of Auditory Regimes of Yugoslav *poena insularis*," a lecture at The Impossible School, Multimedia Institute, Zagreb, March 12, 2020. Author in attendance.

15 Fred L. Griffin, *Creative Listening and the Psychoanalytic Process: Sensibility, Engagement and Envisioning* (New York: Routledge, 2016), 9.

16 Ibid.

17 Nataša Petrešin-Bachelez, "For Slow Institutions," in e-flux journal #85. October 2017; also, Isabelle Stengers, *Another Science Is Possible: A Manifesto for Slow Science* (Cambridge: Polity Press, 2017).
18 Judith Butler, *The Force of Nonviolence: An Ethico-Political Bind* (London: Verso, 2020).
19 See Barry Blesser and Linda-Ruth Salter, *Spaces Speak, Are You Listening? Experiencing Aural Architecture* (Cambridge, MA: The MIT Press, 2006).
20 For more on acoustic ecology, see R. Murray Schafer, *The Soundscape: Our Sonic Environment and the Tuning of the World* (Rochester, VT: Destiny Books, 1994).
21 For more see Jean-François Augoyard and Henry Torgue, *Sonic Experience: A Guide to Everyday Sounds* (Montreal: McGill-Queen's University Press, 2005).
22 Silvia Federici, *Beyond the Periphery of the Skin: Rethinking, Remaking, and Reclaiming the Body in Contemporary Capitalism* (Oakland: PM Press, 2020), 120.
23 Ibid., 123.
24 Ibid., 124.
25 See George P. Fletcher, *Basic Concepts of Legal Thought* (Oxford: Oxford University Press, 1996). For more on theories of justice, and the notion of a just society, see John Rawls, *A Theory of Justice* (Cambridge: Harvard University Press, 1975), Amartya Sen, *The Idea of Justice* (London: Penguin, 2009), Nancy Fraser and Axel Honneth, *Redistribution or Recognition? A Political-Philosophical Exchange* (London: Verso, 2004).
26 Papadopoulos, *Experimental Practice.*
27 Ibid., 110–11.
28 Ibid., 112.
29 See Andreas Philippopoulos-Mihalopoulos, *Spatial Justice: Body, Lawscape, Atmosphere* (London: Routledge, 2015).
30 Giorgio Agamben, *The Use of Bodies* (Stanford: Stanford University Press, 2016), 232.
31 Judith Butler, in *The Force of Nonviolence*, argues that any critical attempts at addressing issues of equality and inequality must additionally address the institutional, political, and social conditions that impact on how equality can be realized. In short, equality is always shaped by the greater institutional and discursive conditions surrounding human lives. See Butler, *The Force of Nonviolence.*
32 Papadopoulos, *Experimental Practice*, 114.
33 Ibid., 113.
34 Axel Honneth, *The Struggle for Recognition: The Moral Grammar of Social Conflicts* (Cambridge: Polity Press, 1995), 174.
35 Ibid., 11.

36 See Cormac Cullinan, *Wild Law: A Manifesto for Earth Justice*, 2nd ed. (Hartford: Chelsea Green Publishing, 2011).

37 James Parker, "The Soundscape of Justice," in *Griffith Law Review*, 20:4 (2011): 964. See also James E. K. Parker, *Acoustic Jurisprudence: Listening to the Trial of Simon Bikindi* (Oxford: Oxford University Press, 2015).

38 With regard to public discourse, I may point to the establishment of the Fairness doctrine in the United States. Initiated in 1949 by the Federal Communications Commission, the doctrine required that broadcasting companies and other media channels give time for controversial issues, as well as presenting opposing views. This was aimed at keeping in check the power of such companies, and the power of the media, in influencing public opinion, as well as to support more equitable access. The Fairness doctrine articulates the importance in maintaining the public sphere as diverse and informed. The Fairness doctrine was overturned in the late 1980s.

39 For a consideration of acoustic commons, see Ella Finer, *Acoustic Commons and the Wild Life of Sound* (Berlin: Errant Bodies Press, forthcoming).

40 In her book, *Listening Publics*, Kate Lacey proposes "the freedom of listening" as a fundamental act that equally supports the freedom of speech and expression. She develops this argument in relation to questions of the public sphere, and the politics of media culture. See Kate Lacey, *Listening Publics: The Politics and Experience of Listening in the Media Age* (Cambridge: Polity Press, 2013).

41 Elena Loizidou, "Dreams and the Political Subject," in Judith Butler, Zeynep Gambetti, and Leticia Sabsay (eds.), *Vulnerability in Resistance* (Durham and London: Duke University Press, 2016).

42 See Sara Ahmed, *Queer Phenomenology: Orientations, Objects, Others* (Durham: Duke University Press, 2006).

43 See Bracha L. Ettinger, *The Matrixial Borderspace* (Minneapolis: University of Minnesota Press, 2006).

44 For more on the project see the related website: https://manusrecordingproject.com/. Accessed March 20, 2019.

45 I'm using the term D/deafness to designate both those who are legally or physically deaf, and those who culturally identify as Deaf. This is common practice within Deaf studies and communities.

46 Questions of Deafness and recognition find a point of dialogue with what Nancy Fraser terms "parity of participation." The principle of parity of participation argues for an egalitarian social and political system: "The basic idea is that equal autonomy, properly understood, entails the real freedom to participate on a par with others in social life. . . . Thus, participatory parity simply *is* the meaning of equal respect for the equal autonomy of human beings qua social actors." Movements of social justice

draw out these perspectives, capturing the degree to which equal opportunity and access contribute to social recognition. See Fraser and Honneth, *Redistribution or Recognition? A Political-Philosophical Exchange*, 231.

47 Donna Haraway, "The Promises of Monsters: A Regenerative Politics for Inappropriate/d Others," in *The Haraway Reader* (New York: Routledge, 2004), 86–8.

48 Angela Davis, *Freedom Is a Constant Struggle: Ferguson, Palestine, and the Foundations of a Movement* (Chicago: Haymarket Books, 2016).

Chapter 1

1 The "work of reorientation" posed by Sara Ahmed is outlined as the processes or acts that seek to navigate a given normativity, which is often orienting one along particular lines. The work of reorientation is thus the operations one may follow in defining other lines. Ahmed, *Queer Phenomenology*.

2 For more on Rancière's notion of politics, and what he terms "the distribution of the sensible," see Rancière, *The Politics of Aesthetics*.

3 Blesser and Salter, *Spaces Speak, Are You Listening?*, 25.

4 Nina Sun Eidsheim, *Sensing Sound: Singing and Listening as Vibrational Practice* (Durham: Duke University Press, 2015), 64. In his article on the history of concert hall design and acoustics, David Prior suggests that concert halls as spaces of performance historically have both mirrored and shaped "ways of listening," particularly in relation to advances in media technologies, such as radio and home stereos. Cultures of music, and the specific forms of listening they engender, are manifested through their spaces of performance and distribution; from the reverberant spaces of the cathedral to the more modern concert arenas dominated by electronic amplification and finally, to the streaming of music to all types of private enclaves, the performance and distribution of music runs parallel to expressions of social and personal orientations. See David Prior, "A Space for Sound: The Rise, Fall and Fallout of the Concert as a Primary Space for Listening," in Colin Ripley with Marco Polo and Arthur Wrigglesworth (eds.), *In the Place of Sound: Architecture, Music, Acoustics* (Newcastle: Cambridge Scholars Publishing, 2007), 125–34. For more on acoustic architecture see also Emily Thompson, *The Soundscape of Modernity: Architectural Acoustics and the Culture of Listening in America, 1900-1933* (Cambridge, MA: The MIT Press, 2002).

5 Eidsheim, *Sensing Sound*, 69.

6 Ibid., 19.

7 Tim Cresswell, *In Place/Out of Place: Geography, Ideology and Transgression* (Minneapolis, MN: University of Minnesota Press, 1996).

8 One such musical example Melioli cites is that of Giovanni Gabrieli and Biagio Marini, and their Sonatas in Echo (sixteenth century). For more see Matteo Melioli, "Inhabiting Soundscape: Architecture of the Unseen World," in Colin Ripley with Marco Polo and Arthur Wrigglesworth (eds.), *In the Place of Sound: Architecture, Music, Acoustics* (Newcastle: Cambridge Scholars Publishing, 2007), 46–8.

9 For more on notions of performativity in her work, see Judith Butler, *Gender Trouble: Feminism and the Subversion of Identity* (London: Routledge, 1999), as well as Judith Butler, *Bodies that Matter: On the Discursive Limits of Sex* (London: Routledge, 2011).

10 While Butler originally poses "performativity" as a way of addressing questions of "gender" and "sex," I'm interested to position the concept to think more broadly about "orientation" and struggles over recognition. In this regard, I stay focused more on the general mechanism of performativity, and the politics of naming, that "all bodies" come to negotiate or live through.

11 For more on vertigo and its effects, see Thomas Brandt, *Vertigo: Its Multisensory Syndromes*, 2nd ed. (New York: Springer, 2002).

12 It is worth noting here how the field of music therapy equally leads us to a greater perspective onto sound and well-being. In their book, *Music Therapy and Traumatic Brain Injury*, Simon Gilbertson and David Aldridge offer an extremely insightful and probing account of music within therapeutic settings. For more, see Simon Gilbertson and David Aldridge, *Music Therapy and Traumatic Brain Injury: A Light on a Dark Night* (London: Jessica Kingsley Publishers, 2010).

13 Eidsheim, *Sensing Sound*, 20. In a lecture by Maria Puig de la Bellacasa and Dimitris Papadopoulos on the eco-commons, they suggest a shift from "sociability" to "commonability" in an attempt to replace human-oriented sociality and kinship with a posthuman model, where "being-in-common" may extend across human and nonhuman life. Lecture held at the Swamp School, Venice Architecture Biennial, September 2018. Author in attendance.

14 Questions of vibration are at the center of Steve Goodman's work, in particular his book *Sonic Warfare*, in which Goodman maps an "ontology of vibrational force." For Goodman, such vibrational force underpins all forms of audition and auditory experience. See Steve Goodman, *Sonic Warfare: Sound, Affect, and the Ecology of Fear* (Cambridge, MA: The MIT Press, 2010).

15 See Barry Truax, *Acoustic Communication* (Norwood, NJ: Ablex Publishing Corporation, 1994).

16 Ibid., 105.

17 See Roland Barthes, *How to Live Together: Novelistic Simulations of Some Everyday Spaces* (New York: Columbia University Press, 2013), 35.

18 Stavros Stavrides, *Towards the City of Thresholds* (Brooklyn: Common Notions, 2019), 69.
19 The notion of "sonic warfare" is developed and theorized by Steve Goodman. See Goodman, *Sonic Warfare: Sound, Affect, and the Ecology of Fear.*
20 See Melioli, "Inhabiting Soundscape: Architecture of the Unseen World."
21 Blesser and Salter, *Spaces Speak, Are You Listening?*, 15.
22 Ibid., 311.
23 Trevor Wishart, *On Sonic Art* (London: Routledge, 1996), 152.
24 Björn Hellström, *Noise Design: Architectural Modelling and the Aesthetics of Urban Acoustic Space* (Göteborg: Bo Ejeby Förlag, 2003), 79.
25 Schafer, *The Soundscape*, 238.
26 Ibid., 205.
27 Ibid., 258.
28 Ibid.
29 I'm drawing the notion of the "sticky" from Sara Ahmed, in particular her work on "queer phenomenology" which I discuss more fully in Chapter 3. For Ahmed, stickiness is mobilized as an image by which to complicate an understanding of phenomenology as the basis for a smooth intersection between the body and the world around. For Ahmed, such intersections are always already frictional. See Ahmed, *Queer Phenomenology.*
30 The term "acousticking" intentionally echoes Christopher Small's notion of "musicking," which works to underscore music less as an object and more as a social performance. See Christopher Small, *Musicking: The Meanings of Performing and Listening* (Middletown, CT: Wesleyan University Press, 1998).
31 Augoyard and Torgue, *Sonic Experience*, 11.
32 Caroline Claus, *Urban Sound Design* (Warsaw: CCA Ujazdowski Castle, 2015), 4.
33 Haraway, "The Promises of Monsters," 69–70. I may equally capture this sense of proliferation that sonic experience generates by reflecting upon the various attempts within urban and architectural design to work with sound in an applied manner. It has been my experience that attempts to utilize sound in practical ways generally falls short in achieving meaningful or substantial results. Beyond questions of volume, noise, and the application of conventional acoustic design strategies, sound and sonic experience seems to elude application.
34 Papadopoulos, *Experimental Practice.*
35 Ibid., 157.
36 Ibid., 198.
37 For more on the particular project I developed see: http://brandonlabelle.net/free_scene.html and the publication, Alex Arteaga and Raquel Rivera (eds.), *Aurality and Environment* (Madrid: Ministry of Culture, 2017). The participating artists

and performers included Vicente Colomar, Fátima Cué, María Escobar, Antonio Gómez, Annie Pui Ling Lok, Catalina Mahecha, and África Clúa Nieto. The work was filmed by the artist Jara Lopez.

38 See: http://latabacalera.net/about-la-tabacalera/. Accessed November 10, 2019.

39 From a lecture by Stavros Stavrides, Making Futures School, Berlin, September 8, 2019. Author in attendance.

40 Stavros Stavrides, Making Futures School, September 8, 2019.

41 In her call for "reclaiming the body" as a source of knowledge and power, Federici underscores listening's role in being able to resist capitalism's grip. Federici, *Beyond the Periphery of the Skin*, 120.

42 The scholar and artist Ella Finer works through the topic of "acoustic commons," considering not only the circulation of sound within social environments but also how audio archives function as common resources. See Finer, *Acoustic Commons and the Wild Life of Sound.*

43 AbdouMaliq Simone, *Improvised Lives: Rhythms of Endurance in an Urban South* (Cambridge: Polity Press, 2019), 27–8.

44 Ada Colau is an important activist in Barcelona, and was part of initiating a number of citizen movements, including Plataforma de Afectados por la Hipoteca (Platform for People Affected by Mortgages) which was founded in 2009 following the financial and housing crisis. She was elected Mayor of Barcelona in 2015 as part of Barcelona En Comú, a citizen-based movement aimed at more participatory and direct forms of democratic governance.

45 Giorgio Agamben, *Potentialities: Collected Essays in Philosophy* (Stanford: Stanford University Press, 1999), 47.

46 Carla Bergman and Nick Montgomery, *Joyful Militancy: Building Thriving Resistance in Toxic Times* (Chico, CA: AK Press, 2017).

47 Fraser and Honneth, *Redistribution or Recognition?*. See also Nancy Fraser, *Justice Interruptus: Critical Reflections on the "Postsocialist" Condition* (New York: Routledge, 1996).

48 For more on Forensic Architecture in general, see Eyal Weizman, *Forensic Architecture: Violence at the Threshold of Detectability* (New York: Zone Books, 2017).

49 For more on the project see: https://saydnaya.amnesty.org/en/about.html. Accessed November 10, 2019.

50 Lawrence Abu Hamdan, "Saydnaya (The Missing 19db)," in James Parker and Joel Stern (eds.), *Eavesdropping: A Reader* (Wellington, NZ: City Gallery Wellington, 2019), 47–8.

51 Ibid.

52 Ibid., 57.

Chapter 2

1 See Alvin Lucier, *I Am Sitting in a Room*, CD (New York: Lovely Music, 1980).
2 Norie Neumark underscores the technological apparatus of this work, emphasizing how such technology is central to the work, suggesting that what we are hearing is, in fact, a technological voice. See Norie Neumark, *Voicetracks: Attuning to Voice in Media and the Arts* (Cambridge: The MIT Press, 2017), 92–3.
3 Zeynep Bulut, "Planes, Walls, and Bits of Sound: Healing a Voice," in *Music and Politics*, X(2) (Summer 2016): 4.
4 For more on musique concrète, and its history, see Pierre Schaeffer, *In Search of a Concrete Music* (Berkeley: University of California Press, 2012). See also the anthology *Audio Culture* for an overview of these and similar issues. Christoph Cox and Daniel Warner (eds.), *Audio Culture: Readings in Modern Music* (New York: Bloomsbury, 2004).
5 Daphne Oram, *An Individual Note of Music, Sound and Electronics* (London: Anomie Publishing, 2016), 17.
6 For more on the project see the artist's website: http://www.zorkawollny.net/?portfolio=resonance-assembly-composition-for-factory. Accessed November 2019.
7 Bulut, "Planes, Walls, and Bits of Sound: Healing a Voice," 6.
8 Jean-Luc Nancy, *Listening* (New York: Fordham University Press, 2007), 15.
9 Ibid., 14.
10 Ibid., 13.
11 Ibid., 16.
12 Maryanne Amacher, *City-Links* pamphlet (New York: Ludlow 38, 2010), 8. http://ludlow38.org/files/mabooklet.pdf. Accessed October 2019.
13 Ibid. According to the pamphlet, the citation first appeared in *Chicago Style—Chicago Today*, May 10, 1974.
14 Joanna Zylinska also provides an important understanding of "relationality" that integrates a broader spectrum of what constitutes externality beyond the strictly human. See Joanna Zylinska, *The End of Man: A Feminist Counterapocalypse* (Minneapolis: University of Minnesota Press, 2018).
15 See Hartmut Rosa, *Resonance: A Sociology of Our Relationship to the World* (Cambridge: Polity, 2019).
16 Ibid., 197.
17 Fred Moten, from a lecture at the University of Chicago. See Fred Moten, "Hesitant Sociology," University of Chicago, May 3, 2016. https://www.youtube.com/watch?v=jlk4X9tZmt4,. Accessed March 10, 2020.
18 For a critical take on resonance, see Robin James, *The Sonic Episteme: Acoustic Resonance, Neoliberalism, and Biopolitics* (Durham: Duke University Press, 2019).

19 Statement by Jana Winderen, found on the artist's website. See http://www.janawinderen.com/fieldtrips/field_trip_to_the_north_pole_m.html#.XqLC4m6xVTY. Accessed April 22, 2020.

20 Extending from Donna Haraway's theories of ecological entanglements, and cyborg identity, it becomes important to also reflect on the presence of the microphone within Jana Winderen's, and other sound artists' methods; how the microphone performs as an actant, contributing actively to the scene of acoustic relationality. The desire to extend beyond the human found within works of sound art is often reliant on particular technological devices and processes, and therefore the experiences evoked must be appreciated as containing the presence of these objects and elements, currents and energies.

21 For more on Jana Winderen, see the artist's website: http://www.janawinderen.com/

22 In tandem with the degree to which such recordings herald a relational intensity, an extra-relationality that attempts to exceed an anthropocentric world, it must be questioned how such acts of recording define beforehand these captured sounds as *other*. Turning the microphone toward the distant, the foreign, and the nonhuman, may performatively situate our listening, to initiate a listening that hears alterity, and therefore defines a relationship of otherness.

23 For more on the work by John Grzinich, see the artist's website: http://maaheli.ee/main/ archives/5000. Accessed November 2019.

24 Citation from "listening as a means to reconnect with the environment: Interview with John Grzinich" conducted by Aleksandra Jach, The Anthropocene Index, December 23, 2015. http://theanthropoceneindex.com/article/33-Listening-as-a-means-to-reconnect-with-the-environment.-Interview-with-John-Grzinich. Accessed April 22, 2020.

25 For more, see the score for *I Am Sitting in a Room*, in Alvin Lucier, *Reflections: Interviews, Scores, Writings* (Cologne: MusikTexte, 1995).

26 Gaston Bachelard, *The Poetics of Reverie: Childhood, Language, and the Cosmos* (Boston: Beacon Press, 1992), 176.

27 Ibid., 3.

28 Ibid., 6.

29 Ibid., 158.

30 Ibid., 14.

31 Roland Barthes, *The Pleasure of the Text* (New York: Hill and Wang, 1975).

32 Pauline Oliveros, "Quantum Listening: From Practice to Theory (to Practice Practice)," in *MusicWorks* Issue #76 (Spring 2000): 73–91.

33 For an informative explanation of Christina Kubisch's "electrical walks" see the related video: https://vimeo.com/54846163

34 Loizidou, "Dreams and the Political Subject," 133.

35 Ibid., 122–45.
36 For more on the embedded "madness" of language, see Jean-Jacques Lecercle, *The Violence of Language* (London: Routledge, 1990).
37 Pablo Neruda, "One Day Stands Out," in *Residence on Earth* (New York: New Directions, 1973), 97–101.
38 See Paul Gilroy, *The Black Atlantic: Modernity and Double Consciousness* (London: Verso Books, 1993).
39 I am focusing on the notion of "double consciousness," following Paul Gilroy, in order to reflect upon the theme of acoustic relationality, and to unpack a political perspective on listening. In doing so, I'm also bracketing off other aspects essential to understanding double consciousness as put forward by W. E. B. Du Bois. In his *The Souls of Black Folk*, Du Bois presents double consciousness not only as a general state emerging from displacement, and the forceful realities of the slave plantation; additionally, double consciousness tries to capture the ways in which black individuals see themselves through their own eyes as well as the eyes of white people. In short, double consciousness is tensed by recognizing oneself as always already "inferior" in the eyes of dominant white society. See W. E. B. Du Bois, *Writings* (New York: Library of America Publications, 1987).
40 Ibid., 16–17.
41 Ibid., 77–8.
42 Kodwo Eshun, *More Brilliant Than the Sun: Adventures in Sonic Fiction* (London: Quartet, 1998), 6.
43 For more on Satch Hoyt and his project, see https://afrosonicmapping.com/. Accessed November 10, 2019. It is important to note, that in addition to the fieldwork and subsequent musical compositions produced, the project also includes the production of paintings, sculptures, and performances, which are presented in collaboration with cultural institutions, importantly the Haus der Kulturen der Welt, Berlin.
44 Judith Butler and Athena Athanasiou, *Dispossessed: Performative in the Political* (Cambridge: Polity Press, 2013), 80.
45 Ibid., 87.
46 Rosa, *Resonance: A Sociology of Our Relationship to the World*, 196.
47 Ibid., 112.
48 In a related self-produced CD/publication by e-Xplo, the group also points to small conversations and informal exchanges, revealing the larger context and experience. These importantly reflect the more textured dynamics defining the relation between artists invited to the Biennial and the workers who must deal with a system of abuse on a daily level. In their interactions, e-Xplo were not unaware of this power relation, and sought to nurture a deeply sensitive and ethical exchange

(which we might understand by way of the title of the project, "I Love To You," and its reference to Luce Irigaray's work on difference and an ethics of speech). The uncertainties and complexities at play in the project became an occasion for the artists to ask in what way contemporary art participates in global markets even while mobilizing critical acts and discourses aimed at unsettling systems of economic and environmental abuse. For more information see the self-published book and CD, *I Love To You: Workers Voices from the U.A.E.*

49 Heimo Lattner, *A Voice That Once Was in One's Mouth* (Berlin: Errant Bodies Press, 2014), 71.

50 Through what she calls "vibrational practices," Nina Sun Eidsheim unfolds a relation to materiality through the connections and linkages vibration conducts. See Eidsheim, *Sensing Sound.*

51 For more on e-Xplo's particular understanding of the voice, see the self-published book and CD, *I Love To You: Workers Voices from the U.A.E.*

52 For more on the "paradox" of voice, see Mladen Dolar, *A Voice and Nothing More* (Cambridge: The MIT Press, 2006).

53 From an artist statement on the work, found on the artist's website. See: http://www.sarahvanhee.com/iscreamed. Accessed November 10, 2019

54 From the script for the performance by Sarah Vanhee found on the artist's website, see: http://www.sarahvanhee.com/iscreamed. Accessed November 10, 2019.

55 Michael Warner, *Publics and Counterpublics* (Cambridge, MA: The MIT Press/Zone Books, 2005), 74.

56 Such arguments around "social imaginaries" can also be found in Benedict Anderson, *Imagined Communities: Reflections on the Origin and Spread of Nationalism* (London: Verso, 2016).

57 Warner, *Publics and Counterpublics*, 114–15.

58 Evelyne Glennie, *Listen World!* (London: Balestier Press, 2019), 25.

59 In a lecture delivered at the University of Cologne, Achille Mbembe addressed the issue of migration and its implications for European sovereignty. In arguing for a more urgent sense of engagement, Mbembe described the presence of the migrant or refugee as a call to which one must ethically and politically respond; in short, to "have the courage to listen to the stranger at the door." The lecture was titled "Concerning the Right to Mobility." June 19, 2019. Author in attendance.

60 Trinh T. Minh-ha, *Elsewhere, Within Here: Immigration, Refugeeism and the Boundary Event* (New York: Routledge, 2011), 2.

61 At the time of its installation, the site operated as a parking lot for the Columbus Art Institute, and also included two original houses (built in the nineteenth century by German immigrants) protected under local legislation. The two buildings

gave inspiration to the idea of *Audible Dwelling* consisting of two structures. See Learning Site, *Audible Dwelling* (Chicago: Half Letter Press, 2011).

62 Glennie, *Listen World!*, 17.

63 For more on the artist Nathalie Harb and her project see: http://nathalieharb.com/the-silent-room/. Accessed March 10, 2020.

64 Dipesh Chakrabarty, from a keynote lecture at the Anthropocene Campus, Lisbon, January 9, 2020. Author in attendance.

Chapter 3

1 Steven Connor suggests that the "auditory I" can be imaged not as a "point," but rather as "membrane," that is, the self pressed from all sides. Here, the skin becomes a deeply active registration and arena for such auditory imaging. See Steven Connor, "Modern Auditory I," in Roy Porter (ed.), *Rewriting the Self: Histories from the Middle Ages to the Present* (London: Routledge, 1996).

2 Didier Anzieu, *The Skin-Ego* (London: Karnac Books, 2016), 44.

3 Maurice Merleau-Ponty, *The Visible and the Invisible* (Evanston: Northwestern University Press, 1964).

4 Naomi Segal, *Consensuality: Didier Anzieu, Gender and the Sense of Touch* (Amsterdam: Rodopi Press, 2009), 4.

5 Ibid., 5.

6 See Ahmed, *Queer Phenomenology*.

7 Adrienne Rich, "Notes Toward a Politics of Location," in *Blood, Bread, and Poetry: Selected Prose 1979-1985* (New York: W. W. Norton & Company, 2015), 215.

8 Sara Ahmed, *Queer Phenomenology*, 14–15.

9 Ibid., 107.

10 See Anzieu, *The Skin-Ego*, 184.

11 Lia García, a transgender artist working in Mexico City, has developed a series of performative projects that specifically contend with how her voice and body are always already at odds with the heteronormative society around her. In conversation with the author, 2019.

12 Tia DeNora, *Music Asylums: Wellbeing through Music in Everyday Life* (London: Routledge, 2006), 120.

13 Ibid.

14 Ibid.

15 Paul Gilroy, "Between the Blues and the Blues Dance: Some Soundscapes of the Black Atlantic," in Michael Bull and Les Back (eds.), *The Auditory Culture Reader* (Oxford: Berg, 2003), 390.

16 Music is emphasized as one such means for working between "roots" of ethnic identity and "routes" toward navigating the social and political structures of the present. In *The Black Atlantic*, Gilroy mentions the importance of music during his time as a young man: "When I was a child and a young man growing up in London, black music provided me with a means to gain proximity to the sources of feeling from which our local conceptions of blackness were assembled." *The Black Atlantic: Modernity and Double Consciousness* (London: Verso Books, 1993), 109.

17 Sirin Özgün outlines questions of sound, listening, and public space by reflecting upon the scenes of demonstrations and celebrations on the streets of Istanbul. See her dissertation, Emine Sirin Özgün, "The Sounds of Political Actions in the Streets of Istanbul," PhD thesis, Istanbul Technical University, Graduate School of Arts and Social Sciences, September 2012.

18 Monique Wittig, *The Lesbian Body* (Boston: Beacon Press, 1986), 86.

19 Ibid., 121–2.

20 In an article on Monique Wittig, Alice Jardine wrestles with the "surprising" universalism Wittig was to argue for; such universalism meant upsetting a "politics of difference" by overcoming the category of "woman," and in favor of the "power of universalization." As Alice Jardine comments, "Wittig wanted to steal universality back from those who had monopolized it." Universality, for Wittig, is not an easy path; rather, it is the work of literature to "embody" a certain point of view. This is brought forward in *The Lesbian Body*, where Wittig shows us a world already shaped by a lesbian language and position. See Alice Jardine, "Thinking Wittig's Differences: 'Or, Failing That, Invent,'" in *GLQ, A Journal of Lesbian and Gay Studies*, 13:4 (2007): 455–66.

21 Anzieu, *The Skin-Ego*, 114.

22 Monique Wittig, "On the Social Contract," in *The Straight Mind and Other Essays* (Boston: Beacon Press, 1992), 45.

23 Wittig, *The Lesbian Body*, 102.

24 Diane Griffin Crowder, "From the Straight Mind to Queer Theory: Implications for Political Movement," in *GLQ, A Journal of Lesbian and Gay Studies*, 13:4 (2007): 493–4. The question of lesbian togetherness finds a point of intersection with the writings of Luce Irigaray. For instance, in her essay "The Sex Which Is Not One" Irigaray argues that women are defined by a form of "auto-eroticism" that cannot be held by the "scopic economy" of the phallus; rather, woman is on the side of a profound tactility. "As for woman, she touches herself in and of herself without any need for mediation, and before there is any way to distinguish activity from passivity. Woman 'touches herself' all the time, and moreover no one can forbid her to do so, for her genitals are formed of two lips in continuous contact." Through such auto-eroticism, and continuous contact, Irigaray draws out an important

feminist view on the body. See Luce Irigaray, "This Sex Which Is Not One," in *This Sex Which Is Not One* (Ithaca: Cornell University Press, 1985), 24.

25 Seth Clark Silberman, "'I Have Access to Your Glottis': The Fleshy Syntax, Ethical Irony, and Queer Intimacy of Monique Wittig's *Le corps lesbien*," in *GLQ, A Journal of Lesbian and Gay Studies*, 13:4 (2007): 473.

26 Jack Halberstam, *In a Queer Time and Place: Transgender Bodies, Subcultural Lives* (New York: New York University Press, 2005).

27 Ibid., 152.

28 Nina Dragičević, from a lecture given at Sounds, Songs, and Politics, International Ethnomusicological symposium, Institute of Ethnomusicology, University of Ljubljana, August 29–31, 2019. Author in attendance.

29 Wittig, *The Lesbian Body*, 38.

30 Kim Rygiel, *Globalizing Citizenship* (Vancouver: University of British Columbia Press, 2010), 159.

31 Ibid., 12.

32 Ibid., 190.

33 Andreas Bandak and Manpreet K. Janeja, "Introduction: Worth the Wait," in *Ethnographies of Waiting: Doubt, Hope and Uncertainty* (London: Bloomsbury Academic, 2019), 5.

34 Synnøve Bendixsen and Thomas Hylland Eriksen, "Time and the Other: Waiting and Hope among Irregular Migrants," in *Ethnographies of Waiting: Doubt, Hope and Uncertainty*, 92.

35 Bandak and Janeja, "Introduction: Worth the Wait," 22.

36 For more on the Manus Recording project see the related website: https://manusrecordingproject.com/. Accessed March 20, 2020.

37 A statement on the project outlines: "Since 24 July 2018, six men—Farhad Bandesh, Behrouz Boochani, Samad Abdul, Shamindan Kanapathi, Kazem Kazemi and Abdul Aziz Muhamat—have been sending daily ten minute audio recordings to The Ian Potter Museum of Art from Manus Island, where they have been detained by the Australian government for the last five years. The recordings are then played back in the gallery throughout the day. This will continue until 28 October 2018 at the end of which 14 hours of sound will have been produced. They are, in effect, developing an archive of what it sounds like to live in limbo." Information provided by the organizers. See also Parker and Stern (eds.), *Eavesdropping*.

38 The detention center on Manus Island was opened in 2012, after having been previously shut in 2008. In 2013 the Australian government announced that asylum seekers and refugees detained on the Island would not be resettled in Australia. Since its opening there has been around 2,000 detainees, with around 200 having been resettled in the United States as part of swap agreements. The remaining

detainees have been relocated to other centers in Papua New Guinea; some also returned to their home countries as part of agreements with the Australian government, and as of late 2019 around 500 remain distributed across three detention centers.

39 The Australian government over the last ten years has been greatly criticized for its policies in ebbing the flow of asylum seekers, implementing not only offshore detention but also boat turnbacks and interventions at ports in Indonesia where most migrant boats originate from. The Manus Island Regional Processing Centre has also been rife with problems, with many detainees experiencing psychological and mental health issues, and there being many cases of suicides, hunger strikes, and also deaths resulting from conflicts with detention guards. In 2016 the government of Papua New Guinea declared the center to be unconstitutional and it was officially closed in October of that year, resulting in a stand-off between police and over 200 detainees.

40 See DeNora, *Music Asylums*, 97.

41 Eidsheim, *Sensing Sound*.

42 Kate Briggs, "Preface," in Roland Barthes, *How to Live Together: Novelistic Simulations of Some Everyday Spaces* (New York: Columbia University Press, 2013), xxii.

43 Henri Lefebvre, *The Production of Space* (Hoboken: Wiley-Blackwell, 1991).

44 Barthes, *How to Live Together*, 35.

45 Ibid.

46 Henri Lefebvre, *Rhythmanalysis: Space, Time and Everyday Life* (London: Bloomsbury, 2013), 15.

47 Ibid., 36.

48 In a related transcript from voice messages sent from the detainees to the organizers of the exhibition in Melbourne, Kazem, a detainee on Manus Island, stated: "You know, Jon . . . today I wanted to record someone. When I talked to them about the reason, and he said to me, Oh, no no, nothing gonna change . . . noone can help us. I did that before, long time ago, many times, and no, nothing is working for us. And after that I told him no problem, brother it's okay—it's not force. If you're happy, we do that. If you're not, no problem, okay. [laughs] And that was my, you know, my day. For recording." Parker and Stern (eds.), *Eavesdropping*, 190.

49 A text message from the detainee Farhad to Jon Tjhia, October 17, 2018.

50 In *The Force of Nonviolence*, Judith Butler elaborates questions of equality and violence through the concept of "grievability." For Butler, equality can be understood, or posed, according to the degree to which certain lives are deemed "grievable," that is, whose loss would be mourned. Grievability comes to function

as an operative term and mechanism in social and political struggles for equality. Judith Butler, *The Force of Nonviolence: An Ethico-Political Bind* (London: Verso, 2020).

51 See Ettinger, *The Matrixial Borderspace.*

52 Ibid., 71.

53 Ibid., 70.

54 Ibid., 75.

55 See Engin F. Isin, "Doing Rights with Things: The Art of Becoming Citizens," in Paula Hildebrandt, Kerstin Evert, Sibylle Peters, Mirjam Schaub, Kathrin Wildner, and Gesa Ziemer (eds.), *Performing Citizenship: Bodies, Agencies, Limitations* (Cham: Palgrave Macmillan), 45–56. See also Angharad Closs Stephens and Vicki Squire, "Politics through a Web: Citizenship and Community Unbound," in *Environment and Planning D: Society and Space*, 30 (2012): 551–67.

56 Behrouz Boochani, *Letter from Manus Island* (Borderstream Books, 2018), 17.

57 Behrouz Boochani, *No Friend but the Mountains* (London: Picador, 2019), 133.

58 See Butler, *The Force of Nonviolence: An Ethico-Political Bind.*

59 In *The Black Atlantic* Paul Gilroy draws out the image of the "fractal" to capture the complex "double reality" of the black Atlantic, and the resulting cultural mixes and emergent metaphysics of black identity. The "fractal" becomes suggestive for underscoring the "mingling of skins" I'm theorizing as cultural and communal productions or relations that remain tensed by the power of the mix. See Gilroy, *The Black Atlantic*, 88.

60 Jean-Luc Nancy, *The Inoperative Community* (Minneapolis: University of Minnesota Press, 1991).

61 Ibid., 61.

62 Ibid., 33.

63 Wittig, *The Lesbian Body*, 7.

64 Monique Wittig, "The Mark of Gender," in *The Straight Mind and Other Essays* (Boston: Beacon Press, 1992), 87.

65 Wittig, *The Lesbian Body*, 109.

66 See Stavros Stavrides, *Common Space: The City as Commons* (London: Zed Books, 2016).

67 I may also point to the pedagogical initiative, Autonomia Akadimia, led by the artist Joulia Strauss, which started in Athens around 2015 and contributes to the general movements of social solidarity. Autonomia Akadimia focuses on bringing people together to share a range of knowledges, to nurture a cooperative culture that can impact onto how we approach being an active citizen. At its core, is a concern for fostering horizontality, bodily awareness, and an expression of radical openness. For an interesting project on Athens and the related solidarity

movements, see also Robin Vanbesien, *Solidarity Poiesis: I Will Come and Steal You* (Berlin: b_books, 2017).

68 Stavrides, *Towards the City of Thresholds*, 38.

69 Ibid., 33.

70 Stavros Stavrides, "Occupied Squares, Societies in Movement," Zed Books website blog. See: https://www.zedbooks.net/blog/posts/occupied-squares-societies-movement/. Accessed April 22, 2020.

71 Stavrides, *Towards the City of Thresholds*, 2.

Chapter 4

1 Lilia Mestre, a choreographer and performance artist based in Brussels. In conversation with the artist, 2019.

2 Gemma Corradi Fiumara, *The Other Side of Language: A Philosophy of Listening* (London: Routledge, 1995), 186.

3 Anna Tsing, in mapping out her notion of "the art of noticing," explicitly draws upon "listening" as a vehicle for enhancing our sense for planetary life. See Anna Lowenhaupt Tsing, *The Mushroom at the End of the World: On the Possibility of Life in Capitalist Ruins* (Princeton: Princeton University Press, 2015).

4 See Roshanak Kheshti, *Modernity's Ear: Listening to Race and Gender in World Music* (New York: New York University Press, 2015).

5 Fiumara, *The Other Side of Language*, 189.

6 Péter Molnár, in conversation with the author, January 2020. Molnár has developed various pedagogical and social initiatives in Budapest, in order to challenge forms of "hate speech," involving people in publicly addressing experiences of discrimination. For more on the Roma in Hungary, see Barbora Černušáková, "The Roma People's Hungarian Hell," in Politico, January 25, 2017. https://www.politico.eu/article/the-roma-peoples-hungarian-hell/. Accessed February 2020. See also the report on Hungary from Minority Rights Group International: https://minorityrights.org/minorities/roma-8/. Accessed February 10, 2020.

7 Raymond Antrobus, *The Perseverance* (London: Penned in the Margins, 2018).

8 H-Dirksen L. Bauman, "Audism: Exploring the Metaphysics of Oppression," in *Journal of Deaf Studies and Deaf Education*, 9:2 (Spring 2004): 240.

9 I'm using the term D/deafness to designate both those who are legally or physically deaf, and those who culturally identify as Deaf. This is common practice within Deaf studies and communities.

10 David Wright, *Deafness: A Personal Account* (London: Faber and Faber, 1991), 9.

11 Ibid., 10–11.
12 Aaron Williamson, "Performance, Disability, Humour and Being in Public: Aaron Williamson in conversation with Marquard Smith," in *Aaron Williamson: Performance, Video, Collaboration* (London: Live Art Development Agency, Kiosk, 2008), 6.
13 Willy Conley, *Listening Through the Bone: Collected Poems* (Washington, DC: Gallaudet University Press, 2018).
14 Paddy Ladd, *Understanding Deaf Culture: In Search of Deafhood* (Clevedon: Multilingual Matters, 2003), 114.
15 For a general resource on current discussions and research, see Journal of Deaf Studies and Deaf Education, published by Oxford University Press, as well as publications from Gallaudet University.
16 H-Dirksen L. Bauman, "Introduction: Listening to Deaf Studies," in *Open Your Eyes: Deaf Studies Talking* (Minneapolis: University of Minnesota Press, 2008), 4.
17 Ibid., 3–4. See also Benjamin Bahan, "Upon the Formation of a Visual Variety of the Human Race," in Ibid., 83–99.
18 H-Dirksen L. Bauman and Joseph J. Murray, "Deaf Gain: An Introduction," in *Deaf Gain: Raising the Stakes on Human Diversity* (Minneapolis: University of Minnesota Press, 2014), xxvi.
19 For more on the question of intimacy, see Teresa Blankmeyer Burke, "Armchairs and Stares: On the Privation of Deafness," in *Deaf Gain*, 11.
20 See David McNeill, *Gesture and Thought* (Chicago: University of Chicago Press, 2007).
21 Bahan, "Upon the Formation of a Visual Variety of the Human Race," 85.
22 For more on William C. Stokoe, see his paper, "Sign Language Structure: An Outline of the Visual Communication Systems of the American Deaf.," originally published in 1960, and reprinted in *Journal of Deaf Studies and Deaf Education*, 10:1 (Winter 2005): 3–37.
23 For more on the importance of video recording within Deaf culture see Marlon Kuntze, "Turning Literacy Inside Out," in *Open Your Eyes*, 146–57.
24 Ladd, *Understanding Deaf Culture: In Search of Deafhood*.
25 Michele Friedner and Stefan Helmreich make a similar proposal in their paper "Sound Studies Meet Deaf Studies," which works at mapping out productive points of contact across the hearing and non-hearing. Michele Friedner and Stefan Helmreich, "Sound Studies Meet Deaf Studies," in *The Senses & Society*, 7:1 (2012): 72–86.
26 Masschelein, "E-ducating the Gaze: The Idea of a Poor Pedagogy" 48.
27 See Sigmund Freud, *Beyond the Pleasure Principle: And Other Writings* (London: Penguin, 2003).

28 Stephen N. Malloch, "Mothers and Infants and Communicative Musicality," in *Musicae Scientiae*, 3 (1999–2000): 32.

29 Ibid., 45.

30 See John Bowlby, *Attachment: Attachment and Loss*, 1 (London: Pimlico, 1997).

31 Hansel Bauman, "DeafSpace: An Architecture toward a More Livable and Sustainable World," in *Deaf Gain*, 379.

32 Wright, *Deafness*, 11–12.

33 For more see Richard E. Cytowic and David M. Eagleman, *Wednesday Is Indigo Blue: Discovering the Brain of Synesthesia* (Cambridge: The MIT Press, 2011).

34 Jane Dillehay, "Genetic Research, Bioethical Issues, and Cochlear Implants: An Overview of the Issues Affecting the Deaf Community," in *Cochlear Implants: Evolving Perspectives* (Washington, DC: Gallaudet University Press, 2011), 29.

35 Masschelein, "E-ducating the Gaze: The Idea of a Poor Pedagogy."

36 Honneth, *The Struggle for Recognition*.

37 Jacques Lacan would emphasize this dynamic of "the gaze" as the realization of one's own lack: the gaze returns to oneself in such a way as to remind of what one cannot have. For more see Jacques Lacan, *The Four Fundamental Concepts of Psychoanalysis* (New York: W. W. Norton and Company, 1998).

38 See Hannah Arendt, *The Human Condition*, second edition (Chicago: University of Chicago Press, 2018).

39 While I'm emphasizing "the gaze" as something that nurtures acknowledgment and recognition, it is clear that the gaze may equally be the gaze of violence, of capture and control: "I see you" is never so clearly benevolent or kind, rather it may also "recognize" in order to hinder or regulate.

40 H-Dirksen L. Bauman, "On the Disconstruction of (Sign) Language in the Western Tradition: A Deaf Reading of Plato's *Cratylus*," in *Open Your Eyes*, 135.

41 Jim Cohn, "The New Deaf Poetics: Visible Poetry," in *Sign Language Studies*, 52 (Fall 1986): 269.

42 Wright, *Deafness*, 11.

43 Oliver Sacks, *Seeing Voices: A Journey into the World of the Deaf* (London: Picador, 2012), 70.

44 Ibid., 85.

45 For more on the expressive force of iconicity, see Cohn, "The New Deaf Poetics: Visible Poetry," 272.

46 Bauman, "Introduction: Listening to Deaf Studies," 18.

47 Bernard Bragg, an important Deaf poet and performer, was a central figure in establishing the National Theater of the Deaf in the United States, in 1967.

48 Sacks, *Seeing Voices*, 78.

49 Maurice Merleau-Ponty, *The World of Perception* (London: Routledge, 2004), 69.

50 Ibid., 70.

51 Julia Kristeva, *Revolution in Poetic Language* (New York: Columbia University Press, 1984), 79.

52 Ibid., 33.

53 Ibid., 99.

54 Cohn, "The New Deaf Poetics: Visible Poetry," 263–77.
Click here to enter text.

55 Ibid., 269.

56 W. J. T. Mitchell, "Preface: Utopian Gestures," in *Signing the Body Poetic: Essays on American Sign Language Literature* (Berkeley: University of California Press, 2006), xvii.

57 See Annette le Fort (ed.), *Parole #2: Phonetic Skin* (Cologne: Salon Verlag, 2012).

58 See Flying Words Project, "Poetry," on accompanying DVD in *Signing the Body Poetic: Essays on American Sign Language Literature.*

59 Adriana Cavarero, *For More than One Voice: Towards a Philosophy of Vocal Expression* (Stanford: Stanford University Press, 2005), 209.

60 McNeill, *Gesture and Thought*, 56.

61 See Dolar, *A Voice and Nothing More.*

62 Miles, "Language for the Eye," cited in Rachel Sutton-Spence, *Analysing Sign Language Poetry* (London: Palgrave, 2004), 243–4.

63 This embodied quality of Sign dynamically expressed within ASL poetry, where the body-sign relation is actively drawn out, and where voice is often also critically explored. For instance, Peter Cook's and Kenny Lerner's duo Flying Words Project perform their poetry together and, by drawing from traditions of mime and Sign, often give expression to a notion of voice that is radically collaborative. Their performance at UCSD in 2008, for example, demonstrates, through a witty and impactful poetics, how voice may also be shaped between two. Here they also playfully explore ventriloquism as Kenny, who is a hearing poet, articulates the spoken voice of Peter, who is Deaf, as he gesticulates and embodies Kenny's voice. Such an approach allows them to put into question the nature of voice, and understandings of deafness, upsetting oralist and audist positions by showing voice as having multiple positions and modalities. See the related video document: https://www.youtube.com/watch?v=rm0WOZqItHg. Accessed April 10, 2020.

64 H-Dirksen L. Bauman, "Getting out of Line: Toward a Visual and Cinematic Poetics of ASL," in *Signing the Body Poetic*, 104.

65 Ibid., 107.

66 Ibid., 109.

67 Tim Ingold, *The Life of Lines* (London: Routledge, 2015), 117.

68 Ibid., 125.

69 Ibid., 128.
70 Ibid., 132.
71 Jim Cohn, *Sign Mind: Studies in American Sign Language Poetics* (Boulder: Museum of American Poetics Publications, 1999), 22.
72 Ibid., 23.
73 Erin Manning, *Politics of Touch: Sense, Movement, Sovereignty* (Minneapolis: University of Minnesota Press, 2008), 12.
74 Ibid., 13.
75 Ibid., 14.
76 Deaf perception and signing has been observed to impact the neurology of the brain, where the auditory cortex is seen to take over functions of seeing things and feeling touch. For more see Donna Jo Napoli, "A Magic Touch: Deaf Gain and the Benefits of Tactile Sensation," in *Deaf Gain*, 211–32.
77 Ibid., 13.
78 Antrobus, *The Perseverance*, 37.
79 Ibid., 108.
80 Sacks, *Seeing Voices*, 96.
81 Jacobus tenBroek, "The Right to Live in the World: The Disabled in the Law of Torts," *California Law Review*, 54:2 (1966): 841–919. tenBroek's article, and legal scholarship, wielded great influence in establishing civil and welfare rights for the disabled.
82 Cohn, *Sign Mind: Studies in American Sign Language Poetics*.
83 Tiziana Terranova, *Network Culture: Politics for the Information Age* (London: Pluto Press, 2004).
84 I'm reflecting upon a performance work by Aaron Williamson, a Deaf artist based in England and whose works often focus on the tensions between hearing, speech, and deafness. The particular performance work I consider here is titled *Phantom Shifts* (1999). For more see the accompanying DVD in the publication, *Signing the Body Poetic*.
85 José Gil, *Metamorphoses of the Body* (Minneapolis: University of Minnesota Press, 1998), 124.
86 Ibid., 136.
87 Ibid., 170.
88 Franco "Bifo" Berardi, *The Soul at Work: From Alienation to Autonomy* (Los Angeles: Semiotext(e), 2009).
89 For an insightful overview of mindfulness, and questions of attunement and well-being, see Daniel J. Siegel, *The Mindful Brain: Reflection and Attunement in the Cultivation of Well-Being* (New York: W. W. Norton & Company, 2007).
90 For a detailed overview of histories of deafness, see Jonathan Rée, *I See a Voice: Deafness, Language and the Senses—A Philosophical History* (New York: Metropolitan Books, 1999).

91 Ladd, *Understanding Deaf Culture: In Search of Deafhood*, 159.
92 See Raylene Paludneviciene and Raychelle L. Harris, "Impact of Cochlear Implants on the Deaf Community". in *Cochlear Implants: Evolving Perspectives* (Washington, DC: Gallaudet University Press, 2011), 3–19.
93 For more on The Learning Center for the Deaf, see: https://www.tlcdeaf.org/.
94 For an insightful view on experiences of cochlear implant usage, see Michael Chorost, *World Wide Mind*. "My cochlear implant substitutes for the lost hair cells by directly triggering the auditory nerves with implanted electrodes. A surgeon drilled an inch and a half into my skull, countersunk a ceramic-encased micro-chip behind my left ear, and threaded sixteen electrodes into my inner ear. Now an external device sitting on my ear picks up sound, digitizes it, and radios a stream of 1s and 0s through my skin to the microchip. The chip receives the radio signal with a tiny antenna and decides how to strobe the electrodes on and off. By choosing which electrodes to fire at any given moment, it makes my auditory nerves transmit sound information to my brain.
Even though I have 280,000 transistors in my skull, more than in the CPU of my computer when I started grad school, they can't reproduce the functioning of a normal ear in all its subtlety and range. In fact, they stimulate the auditory nerves in a way that is quite different than in a normal ear. Because of that, I had to learn how to hear all over again. Voices sounded like gibberish at first. It took me months to learn how to interpret the software's representation of vowels and consonants as English." Michael Chorost, *World Wide Mind: The Coming Integration of Humans and Machines* (New York: Free Press, 2016), 6.
95 In reflecting upon cochlear implantation, and its "electroacoustic" politics, Mara Mills suggestively poses the term "deaf futurism." Mara Mills, "Do Signals Have Politics? Inscribing Abilities in Cochlear Implants." In *The Oxford Handbook of Sound Studies*, eds. Trevor Pinch and Karin Bijsterveld (Oxford: Oxford University Press, 2011), 320–46.
96 Paludneviciene and Harris, "Impact of Cochlear Implants on the Deaf Community," 13.
97 Dillehay, "Genetic Research, Bioethical Issues, and Cochlear Implants."
98 Ibid.
99 Mills, "Do Signals Have Politics? Inscribing Abilities in Cochlear Implants." The "politics of signals" that Mills points at can also be appreciated by considering the use of hearing aids, telecommunications relay services, induction loops, as well as FM or infrared transmission. These devices further articulate a relation to "signals" experienced and utilized by the Deaf community.
100 See Tsing, *The Mushroom at the End of the World*, especially chapter 1 where the author addresses "the art of noticing."

101 Jonathan Sterne, *The Audible Past: Cultural Origins of Sound Reproduction* (Durham: Duke University Press, 2003), 347.

102 See Lautstrom #36: Social Currency or Skin? Notions on Voice, Absence and Presence of Sound: A conversation with Christine Sun Kim and Zeynep Bulut, conceived and facilitated by Anna Bromley, September 25, 2013. Broadcast on reboot.fm. December 2013. https://www.annabromley.com/readlautstrom. Accessed March 2020.

103 See the TED talk by Christine Sun Kim, "The enchanting music of sign langauge." https://www.youtube.com/watch?v=2Euof4PnjDk. Accessed April 25, 2020.

104 For more on the project, *Infinite Ear*, see the related website: http://www.council.art/inquiries/30/infinite-ear. Accessed March 20, 2020.

105 "Surdo, Logo Existo" is a theater work developed, directed, and performed by Deaf artists, staged at Teatro Guaíra, Curitiba, September 2019. Direction: Gabriela Grigolom and Jonathas Medeiros. Produced by Chiris Gomes, Viviana Medeiros, and Felipe Patrício. The photographs presented throughout this chapter capture a scene of the play, which takes place in an imaginary location, in 1880, date of the prohibition of the use of sign language in European schools known as "the Milan Conference." After deliberations, the conference declared that oral education was superior to manual education and passed a resolution banning the use of sign language in school. This resolution changed only in the 1960s when the minoritarian movements, such as feminism, racial, and environmental movements, made it possible for the Deaf communities to vindicate their cultural and linguistic rights and to officially use sign language as their first language. For more on Grupo Mapas, and the theater work "Surdo, logo existo!," see the website: https://olharcomum.wordpress.com/2019/09/26/surdo-logo-existo-teatro-guaira-mini-auditorio/

106 Tiqqun, *Introduction to Civil War* (Los Angeles: Semiotext(e), 2010), 16.

107 The term "deaf gain" is put forward in the publication of the same title in order to capture the positive contributions or productive paths opened up by way of Deaf culture. See *Deaf Gain: Raising the Stakes on Human Diversity*.

Acoustic Support

1 I draw particular guidance through the writings of Paul Gilroy and Fred Moten, as well as many poetic and musical expressions that speak toward the black Atlantic, and the cultural identities of black community life. For more see Gilroy, *The Black Atlantic* and Fred Moten, *In the Break: The Aesthetics of the Black Radical Tradition* (Minneapolis: University of Minnesota Press, 2003).

2 See Anzieu, *The Skin-Ego*.

3 Balibar, *Violence and Civility*.

Bibliography

Agamben, Giorgio. *Potentialities: Collected Essays in Philosophy*. Stanford: Stanford University Press, 1999.

Agamben, Giorgio. *The Use of Bodies*. Stanford: Stanford University Press, 2016.

Ahmed, Sara. *Queer Phenomenology: Orientations, Objects, Others*. Durham: Duke University Press, 2006.

Amacher, Maryanne. *City-Links* pamphlet. New York: Ludlow 38, 2010. Accessed October 1, 2019. http:// ludlow38.org/files/mabooklet.pdf.

Anderson, Benedict. *Imagined Communities: Reflections on the Origin and Spread of Nationalism*. London: Verso, 2016.

Antrobus, Raymond. *The Perseverance*. London: Penned in the Margins, 2018.

Anzieu, Didier. *The Skin-Ego*. London: Karnac Books, 2016.

Arendt, Hannah. *The Human Condition*, second edition. Chicago: University of Chicago Press, 2018.

Arteaga, Alex and Rivera, Raquel, eds. *Aurality and Environment*. Madrid: Ministry of Culture, 2017.

Augoyard, Jean-François and Torgue, Henry. *Sonic Experience: A Guide to Everyday Sounds*. Montreal: McGill-Queen's University Press, 2005.

Azhari, Timour. "Defiant Protestors Hold Rival Parade on Lebanon Independence Day." In Al Jazeera, November 23, 2019. Accessed March 10, 2020. https://www.aljazeera.com/news/2019/11/defiant-protesters-hold-rival-parade-lebanon-independence-day-191122202937296.html.

Bachelard, Gaston. *The Poetics of Reverie: Childhood, Language, and the Cosmos*. Boston: Beacon Press, 1992.

Bahan, Benjamin. "Upon the Formation of a Visual Variety of the Human Race." In *Open Your Eyes: Deaf Studies Talking*, edited by H-Dirksen L. Bauman, 83–99. Minneapolis: University of Minnesota Press, 2008.

Balibar, Étienne. *Violence and Civility: On the Limits of Political Philosophy*. New York: Columbia University Press, 2016.

Bandak, Andreas and Janeja, Manpreet K. "Introduction: Worth the Wait." In *Ethnographies of Waiting: Doubt, Hope and Uncertainty*, edited by Andreas Bandak and Manpreet K. Janeja, 3–41. London: Bloomsbury Academic, 2019.

Barthes, Roland. *The Pleasure of the Text*. New York: Hill and Wang, 1975.

Barthes, Roland. *The Responsibility of Forms: Critical Essays on Music, Art, and Representation*. Berkeley: University of California Press, 1991.

Barthes, Roland. *How to Live Together: Novelistic Simulations of Some Everyday Spaces*. New York: Columbia University Press, 2013.

Battah, Habib. "An Open Mic Revolution in Lebanon." In Al Jazeera, October 25, 2019. Accessed March 10, 2020. https://www.aljazeera.com/indepth/opinion/open-mic-revolution-lebanon-191025150455130.html.

Battah, Habib. "A New Politics Is Rising in Lebanon." In Al Jazeera, November 15, 2019. Accessed March 10, 2020. https://www.aljazeera.com/indepth/features/politics-rising-lebanon-191114141631074.html

Bauman, H-Dirksen L. "Audism: Exploring the Metaphysics of Oppression." In *Journal of Deaf Studies and Deaf Education*, 9:2 (Spring 2004): 239–46.

Bauman, H-Dirksen L. "Introduction: Listening to Deaf Studies." In *Open Your Eyes: Deaf Studies Talking*, edited by H-Dirksen L. Bauman, 1–34. Minneapolis: University of Minnesota Press, 2008.

Bauman, H-Dirksen L. "Getting Out of Line: Toward a Visual and Cinematic Poetics of ASL." In *Signing the Body Poetic: Essays on American Sign Language Literature*, edited by H-Dirksen L. Bauman, Jennifer L. Nelson, and Heidi M. Rose, 95–117. Berkeley: University of California Press, 2006.

Bauman, H-Dirksen L. "On the Disconstruction of (Sign) Language in the Western Tradition: A Deaf Reading of Plato's *Cratylus*." In *Open Your Eyes: Deaf Studies Talking*, edited by H-Dirksen L. Bauman, 83–99. Minneapolis: University of Minnesota Press, 2008.

Bauman, H-Dirksen L. and Murray, Joseph J. "Deaf Gain: An Introduction." In *Deaf Gain: Raising the Stakes on Human Diversity*, edited by H-Dirksen L. Bauman and Joseph J. Murray, xv–xlii. Minneapolis: University of Minnesota Press, 2014.

Bauman, Hansel. "DeafSpace: An Architecture Toward a More Livable and Sustainable World." In *Deaf Gain: Raising the Stakes on Human Diversity*, edited by H-Dirksen L. Bauman and Joseph J. Murray, 375–401. Minneapolis: University of Minnesota Press, 2014.

Bendixsen, Synnøve and Eriksen, Thomas Hylland. "Time and the Other: Waiting and Hope among Irregular Migrants." In *Ethnographies of Waiting: Doubt, Hope and Uncertainty*, edited by Andreas Bandak and Manpreet K. Janeja, 87–112. London: Bloomsbury Academic, 2019.

Berardi, Franco. "Bifo." *The Soul at Work: From Alienation to Autonomy*. Los Angeles: Semiotext(e), 2009.

Bergman, Carla and Montgomery, Nick. *Joyful Militancy: Building Thriving Resistance in Toxic Times*. Chico, CA: AK Press, 2017.

Blesser, Barry and Salter, Linda-Ruth. *Spaces Speak, Are You Listening? Experiencing Aural Architecture*. Cambridge, MA: The MIT Press, 2006.

Boochani, Behrouz. *Letter from Manus Island*. Australia: Borderstream Books, 2018.

Boochani, Behrouz. *No Friend but the Mountains*. London: Picador, 2019.

Bowlby, John. *Attachment: Attachment and Loss*, Vol. 1. London: Pimlico, 1997.

Brandt, Thomas. *Vertigo: Its Multisensory Syndromes*, 2nd edition. New York: Springer, 2002.

Bulut, Zeynep. "Planes, Walls, and Bits of Sound: Healing a Voice." *Music and Politics*, X:2 (2016): 1–16.

Burke, Teresa Blankmeyer. "Armchairs and Stares: On the Privation of Deafness." In *Deaf Gain: Raising the Stakes on Human Diversity*, edited by H-Dirksen L. Bauman and Joseph J. Murray, 3–22. Minneapolis: University of Minnesota Press, 2014.

Butler, Judith. *Gender Trouble: Feminism and the Subversion of Identity*. London: Routledge, 1999.

Butler, Judith. *Bodies that Matter: On the Discursive Limits of Sex*. London: Routledge, 2011.

Butler, Judith. *The Force of Nonviolence: An Ethico-Political Bind*. London: Verso, 2020.

Butler, Judith and Athanasiou, Athena. *Dispossessed: Performative in the Political*. Cambridge: Polity Books, 2013.

Cavarero, Adriana. *For More than One Voice: Towards a Philosophy of Vocal Expression*. Stanford: Stanford University Press, 2005.

Černušáková, Barbora. "The Roma People's Hungarian Hell." Politico, January 25, 2017. Accessed February 5, 2020. https://www.politico.eu/article/the-roma-peoples-hungarian-hell/.

Chorost, Michael. *World Wide Mind: The Coming Integration of Humans and Machines*. New York: Free Press, 2016.

Claus, Caroline. *Urban Sound Design*. Warsaw: CCA Ujazdowski Castle, 2015.

Cohn, Jim. "The New Deaf Poetics: Visible Poetry." *Sign Language Studies*, 52 (Fall 1986): 263–77. Accessed April 8, 2020. www.jstor.org/stable/26203912.

Cohn, Jim. *Sign Mind: Studies in American Sign Language Poetics*. Boulder: Museum of American Poetics Publications, 1999.

Conley, Willy. *Listening through the Bone: Collected Poems*. Washington, DC: Gallaudet University Press, 2018.

Connor, Steven. "The Modern Auditory I." In *Rewriting the Self: Histories from the Middle Ages to the Present*, edited by Roy Porter, 203–23. London: Routledge, 1996.

Connor, Steven. "Writing the White Voice." A talk given at *Sound, Silence and the Arts* symposium, Nanyang Technological University, Singapore, February 28, 2009. Found on the author's website: Accessed March 10, 2020. http://stevenconnor.com/whitevoice.html.

Cox, Christoph and Warner, Daniel, eds. *Audio Culture: Readings in Modern Music*. New York: Bloomsbury, 2004.

Cresswell, Tim. *In Place/Out of Place: Geography, Ideology and Transgression*. Minneapolis, MN: University of Minnesota Press, 1996.

Crowder, Diane Griffin. "From the Straight Mind to Queer Theory: Implications for Political Movement." *GLQ, A Journal of Lesbian and Gay Studies*, 13:4 (2007): 489–503.

Cullinan, Cormac. *Wild Law: A Manifesto for Earth Justice*, second edition. Hartford: Chelsea Green Publishing, 2011.

Cytowic, Richard E. and Eagleman, David M. *Wednesday Is Indigo Blue: Discovering the Brain of Synesthesia*. Cambridge: The MIT Press, 2011.

Davis, Angela Y. *Freedom Is a Constant Struggle: Ferguson, Palestine, and the Foundations of a Movement*. Chicago: Haymarket Books, 2016.

DeNora, Tia. *Music Asylums: Wellbeing Through Music in Everyday Life*. London: Routledge, 2006.

Devisch, Ignaas. *Jean-Luc Nancy and the Question of Community*. London: Bloomsbury, 2013.

Dillehay, Jane. "Genetic Research, Bioethical Issues, and Cochlear Implants: An Overview of the Issues Affecting the Deaf Community." In *Cochlear Implants: Evolving Perspectives*, edited by Raylene Paludneviciene and Irene W. Leigh, 20–38. Washington, DC: Gallaudet University Press, 2011.

Dolar, Mladen. *A Voice and Nothing More*. Cambridge, MA: The MIT Press, 2006.

Du Bois, W. E. B. *Writings*. New York: Library of America Publications, 1987.

Eidsheim, Nina Sun. *Sensing Sound: Singing and Listening as Vibrational Practice*. Durham: Duke University Press, 2015.

Eidsheim, Nina Sun. *The Race of Sound: Listening, Timbre and Vocality in African American Music*. Durham: Duke University Press, 2019.

Eshun, Kodwo. *More Brilliant than the Sun: Adventures in Sonic Fiction*. London: Quartet, 1998.

Ettinger, Bracha L. *The Matrixial Borderspace*. Minneapolis: University of Minnesota Press, 2006.

Farinati, Lucia and Firth, Claudia. *The Force of Listening*. Berlin: Errant Bodies Press, 2017.

Federici, Silvia. *Beyond the Periphery of the Skin: Rethinking, Remaking, and Reclaiming the Body in Contemporary Capitalism*. Oakland: PM Press, 2020.

Finer, Ella. *Acoustic Commons and the Wild Life of Sound*. Berlin: Errant Bodies Press, forthcoming.

Fiumara, Gemma Corradi. *The Other Side of Language: A Philosophy of Listening*. London: Routledge, 1995.

Fletcher, George P. *Basic Concepts of Legal Thought*. Oxford: Oxford University Press, 1996.

Flying Words Project, "Poetry." Accompanying DVD in *Signing the Body Poetic: Essays on American Sign Language Literature*, edited by H-Dirksen L. Bauman, Jennifer L. Nelson, and Heidi M. Rose. Berkeley: University of California Press, 2006.

Fraser, Nancy. *Justice Interruptus: Critical Reflections on the "Postsocialist" Condition*. New York: Routledge, 1996.

Fraser, Nancy and Honneth, Axel. *Redistribution or Recognition? A Political-Philosophical Exchange*. London: Verso, 2004.

Freud, Sigmund. *Beyond the Pleasure Principle: And Other Writings*. London: Penguin, 2003.

Friedner, Michele and Helmreich, Stefan. "Sound Studies Meet Deaf Studies." *The Senses & Society*, 7:1 (2012): 72–86.

Gil, José. *Metamorphoses of the Body*. Minneapolis: University of Minnesota Press, 1998.

Gilbertson, Simon and Aldridge, David. *Music Therapy and Traumatic Brain Injury: A Light on a Dark Night*. London: Jessica Kingsley Publishers, 2010.

Gilroy, Paul. *The Black Atlantic: Modernity and Double Consciousness*. London: Verso Books, 1993.

Gilroy, Paul. "Between the Blues and the Blues Dance: Some Soundscapes of the Black Atlantic." In *The Auditory Culture Reader*, edited by Michael Bull and Les Back, 381–96. Oxford: Berg, 2003.

Glennie, Evelyne. *Listen World!* London: Balestier Press, 2019.

Goodman, Steve. *Sonic Warfare: Sound, Affect, and the Ecology of Fear*. Cambridge, MA: The MIT Press, 2010.

Griffin, Fred L. *Creative Listening and the Psychoanalytic Process: Sensibility, Engagement and Envisioning*. New York: Routledge, 2016.

Halberstam, Jack. *In a Queer Time and Place: Transgender Bodies, Subcultural Lives*. New York: New York University Press, 2005.

Hamdan, Lawrence Abu. "Saydnaya (The Missing 19db)." In *Eavesdropping: A Reader*, edited by James Parker and Joel Stern, 42–59. Wellington, NZ: City Gallery Wellington, 2019.

Haraway, Donna. *The Haraway Reader*. New York: Routledge, 2004.

Harney, Stefano and Moten, Fred. *The Undercommons: Fugitive Planning and Black Study*. New York: Minor Compositions, 2013.

Hartman, Saidiya. "Venus in Two Acts." *Small Axe, Number 26*, 12:2 (2008): 1–14.

Heller-Roazen, Daniel. *The Inner Touch: Archeology of a Sensation*. New York: Zone Books, 2007.

Hellström, Björn. *Noise Design: Architectural Modelling and the Aesthetics of Urban Acoustic Space*. Göteborg: Bo Ejeby Förlag, 2003.

Helmreich, Stefan. *Sounding the Limits of Life: Essays in the Anthropology of Biology and Beyond*. Princeton: Princeton University Press, 2016.

Honneth, Axel. *The Struggle for Recognition: The Moral Grammar of Social Conflicts*. Cambridge: Polity Press, 1995.

Ingold, Tim. *The Life of Lines*. London: Routledge, 2015.

Irigaray, Luce. *This Sex Which Is Not One*. Ithaca: Cornell University Press, 1985.

Isin, Engin F. "Doing Rights with Things: The Art of Becoming Citizens." In *Performing Citizenship: Bodies, Agencies, Limitations*, edited by Paula Hildebrandt, Kerstin Evert, Sibylle Peters, Mirjam Schaub, Kathrin Wildner, and Gesa Ziemer, 45–56. Cham: Palgrave Macmillan, 2019.

Jach, Aleksandra. "Listening as a Means to Reconnect with the Environment: Interview with John Grzinich." The Anthropocene Index, December 23, 2015. Accessed April 23, 2020. http://theanthropoceneindex.com/article/33-Listening-as-a-means-to-reconnect-with-the-environment.-Interview-with-John-Grzinich.

James, Robin. *The Sonic Episteme: Acoustic Resonance, Neoliberalism, and Biopolitics*. Durham: Duke University Press, 2019.

Jardine, Alice. "Thinking Wittig's Differences: 'Or, Failing That, Invent.'" *GLQ, A Journal of Lesbian and Gay Studies*, 13:4 (2007): 455–66.

Kheshti, Roshanak. *Modernity's Ear: Listening to Race and Gender in World Music*. New York: New York University Press, 2015.

Kristeva, Julia. *Revolution in Poetic Language*. New York: Columbia University Press, 1984.

Kuntze, Marlon. "Turning Literacy Inside Out." In *Open Your Eyes: Deaf Studies Talking*, edited by H-Dirksen L. Bauman, 146–57. Minneapolis: University of Minnesota Press, 2008.

Lacan, Jacques. *The Four Fundamental Concepts of Psychoanalysis*. New York: W.W. Norton and Company, 1998.

Lacey, Kate. *Listening Publics: The Politics and Experience of Listening in the Media Age*. Cambridge: Polity Press, 2013.

Ladd, Paddy. *Understanding Deaf Culture: In Search of Deafhood*. Clevedon: Multilingual Matters, 2003.

Lattner, Heimo. *A Voice That Once Was in One's Mouth*. Berlin: Errant Bodies Press, 2014.

Lautstrom #36: "Social Currency or Skin? Notions on Voice, Absence and Presence of Sound: A conversation with Christine Sun Kim and Zeynep Bulut, Conceived and Facilitated by Anna Bromley." September 25, 2013. Broadcast on reboot.fm. December 2013. Accessed March 10, 2020. https://www.annabromley.com/readlau tstrom. Accessed March 2020.

Learning Site. *Audible Dwelling*. Chicago: Half Letter Press, 2011.

Lecercle, Jean-Jacques. *The Violence of Language*. London: Routledge, 1990.

Lefebvre, Henri. *The Production of Space*. Hoboken: Wiley-Blackwell, 1991.

Lefebvre, Henri. *Rhythmanalysis: Space, Time and Everyday Life*. London: Bloomsbury, 2013.

le Fort, Annette, ed. *Parole #2: Phonetic Skin*. Cologne: Salon Verlag, 2012.

Levin, David Michael. *The Listening Self: Personal Growth, Social Change and the Closure of Metaphysics*. London: Routledge, 1989.

Loizidou, Elena. "Dreams and the Political Subject." In *Vulnerability in Resistance*, edited by Judith Butler, Zeynep Gambetti, and Leticia Sabsay, 122–45. Durham: Duke University Press, 2016.

Lucier, Alvin. *I Am Sitting in a Room*, CD. New York: Lovely Music, 1980.

Lucier, Alvin. *Reflections: Interviews, Scores, Writings*. Cologne: MusikTexte, 1995.

Malloch, Stephen N. "Mothers and Infants and Communicative Musicality." *Musicae Scientiae*, 3 (1999–2000): 29–57.

Manning, Erin. *Politics of Touch: Sense, Movement, Sovereignty*. Minneapolis: University of Minnesota Press, 2008.

Masschelein, Jan. "E-ducating the Gaze: The Idea of a Poor Pedagogy." *Ethics and Education*, 5:1 (2010): 43–53.
McNeill, David. *Gesture and Thought*. Chicago: University of Chicago Press, 2007.
Melioli, Matteo. "Inhabiting Soundscape: Architecture of the Unseen World." In *In the Place of Sound: Architecture, Music, Acoustics*, edited by Colin Ripley with Marco Polo and Arthur Wrigglesworth, 46–8. Newcastle: Cambridge Scholars Publishing, 2007.
Merleau-Ponty, Maurice. *The Visible and the Invisible*. Evanston: Northwestern University Press, 1964.
Merleau-Ponty, Maurice. *The World of Perception*. London: Routledge, 2004.
Miles, Dorothy. "Language for the Eye." In Rachel Sutton-Spence, *Analysing Sign Language Poetry*, 243–4. London: Palgrave, 2004.
Mills, Mara. "Do Signals Have Politics? Inscribing Abilities in Cochlear Implants." In *The Oxford Handbook of Sound Studies*, edited by Trevor Pinch and Karin Bijsterveld, 320–46. Oxford: Oxford University Press, 2011.
Minh-ha, Trinh T. *Elsewhere, Within Here: Immigration, Refugeeism and the Boundary Event*. New York: Routledge, 2011.
Mitchell, W. J. T. "Preface: Utopian Gestures." In *Signing the Body Poetic: Essays on American Sign Language Literature*, edited by H-Dirksen L. Bauman, Jennifer L. Nelson, and Heidi M. Rose, xv–xxiv. Berkeley: University of California Press, 2006.
Moten, Fred. *In the Break: The Aesthetics of the Black Radical Tradition*. Minneapolis: University of Minnesota Press, 2003.
Moten, Fred. "Hesitant Sociology." Lecture at University of Chicago, May 3, 2016. Accessed March 15, 2020. https://www.youtube.com/watch?v=jlk4X9tZmt4.
Nancy, Jean-Luc. *The Inoperative Community*. Minneapolis: University of Minnesota Press, 1991.
Nancy, Jean-Luc. *Listening*. New York: Fordham University Press, 2007.
Napoli, Donna Jo. "A Magic Touch: Deaf Gain and the Benefits of Tactile Sensation." In *Deaf Gain: Raising the Stakes on Human Diversity*, edited by H-Dirksen L. Bauman and Joseph J. Murray, 211–32. Minneapolis: University of Minnesota Press, 2014.
Neruda, Pablo. *Residence on Earth*. New York: New Directions, 1973.
Neumark, Norie. *Voicetracks: Attuning to Voice in Media and the Arts*. Cambridge, MA: The MIT Press, 2017.
Oliveros, Pauline. "Quantum Listening: From Practice to Theory (to Practice Practice)." *MusicWorks*, #76 (Spring 2000): 73–91.
Oram, Daphne. *An Individual Note of Music, Sound and Electronics*. London: Anomie Publishing, 2016.
Özgün, Emine Sirin. "The Sounds of Political Actions in the Streets of Istanbul." PhD thesis, Istanbul Technical University, Graduate School of Arts and Social Sciences, September 2012.
Paludneviciene, Raylene and Harris, Raychelle L. "Impact of Cochlear Implants on the Deaf Community." In *Cochlear Implants: Evolving Perspectives*, edited by Raylene

Paludneviciene and Irene W. Leigh, 3–19. Washington, DC: Gallaudet University Press, 2011.
Papadopoulos, Dimitris. *Experimental Practice: Technoscience, Alterontologies, and More-Than-Social Movements*. Durham: Duke University Press, 2018.
Parker, James. "The Soundscape of Justice." *Griffith Law Review*, 20:4 (2011): 965–93.
Parker, James E. K. *Acoustic Jurisprudence: Listening to the Trial of Simon Bikindi*. Oxford: Oxford University Press, 2015.
Parker, James and Stern, Joel, eds. *Eavesdropping: A Reader*. Wellington, NZ: City Gallery Wellington, 2019.
Patchett, Emma and Keenan, Sarah. *Spatial Justice and Diaspora*. Oxford: Counterpress, 2017.
Petrešin-Bachelez, Nataša. "For Slow Institutions." *e-flux journal* #85 (October 2017): 1–13.
Philippopoulos-Mihalopoulos, Andreas. *Spatial Justice: Body, Lawscape, Atmosphere*. London: Routledge, 2015.
Piškor, Mojca. "Reading the Sound, Listening through Words: Resonances of Auditory Regimes of Yugoslav *poena insularis*." Lecture at The Impossible School, Multimedia Institute, Zagreb, March 12, 2020. Author in attendance.
Prior, David. "A Space for Sound: The Rise, Fall and Fallout of the Concert as a Primary Space for Listening." In *In the Place of Sound: Architecture, Music, Acoustics*, edited by Colin Ripley with Marco Polo and Arthur Wrigglesworth, 125–34. Newcastle: Cambridge Scholars Publishing, 2007.
Rancière, Jacques. *The Politics of Aesthetics*. London: Bloomsbury, 2013.
Rawls, John. *A Theory of Justice*. Cambridge: Harvard University Press, 1975.
Rée, Jonathan. *I See a Voice: Deafness, Language and the Senses – A Philosophical History*. New York: Metropolitan Books, 1999.
Rich, Adrienne. "Notes Toward a Politics of Location." In *Blood, Bread, and Poetry: Selected Prose 1979-1985*. New York: W. W. Norton & Company, 2015.
Rosa, Hartmut. *Resonance: A Sociology of Our Relationship to the World*. Cambridge: Polity, 2019.
Rygiel, Kim. *Globalizing Citizenship*. Vancouver: University of British Columbia Press, 2010.
Sacks, Oliver. *Seeing Voices: A Journey into the World of the Deaf*. London: Picador, 2012.
Schaeffer, Pierre. *In Search of a Concrete Music*. Berkeley: University of California Press, 2012.
Schafer, R. Murray. *The Soundscape: Our Sonic Environment and the Tuning of the World*. Rochester, VT: Destiny Books, 1994.
Segal, Naomi. *Consensuality: Didier Anzieu, Gender and the Sense of Touch*. Amsterdam: Rodopi Press, 2009.
Sen, Amataya. *The Idea of Justice*. London: Penguin, 2009.
Siegel, Daniel J. *The Mindful Brain: Reflection and Attunement in the Cultivation of Well-Being*. New York: W.W. Norton & Company, 2007.

Silberman, Seth Clark. " 'I Have Access to Your Glottis': The Fleshy Syntax, Ethical Irony, and Queer Intimacy of Monique Wittig's *Le corps Lesbien*." *GLQ, A Journal of Lesbian and Gay Studies*, 13:4 (2007): 467–88.

Simone, AbdouMaliq. "People as Infrastructure: Intersecting Fragments in Johannesburg." *Public Culture*, 16:3 (2004): 407–29.

Simone, AbdouMaliq. *Improvised Lives: Rhythms of Endurance in an Urban South*. Cambridge: Polity Press, 2019.

Small, Christopher. *Musicking: The Meanings of Performing and Listening*. Middletown, CT: Wesleyan University Press, 1998.

Stavrides, Stavros. *Common Space: The City as Commons*. London: Zed Books, 2016.

Stavrides, Stavros. *Towards the City of Thresholds*. Brooklyn: Common Notions, 2019.

Stavrides, Stavros. "Occupied Squares, Societies in Movement." Zed Books website blog. Accessed April 22, 2020. https://www.zedbooks.net/blog/posts/occupied-squares-societies-movement/.

Stengers, Isabelle. *Another Science Is Possible: A Manifesto for Slow Science*. Cambridge: Polity Press, 2017.

Stephens, Angharad Closs and Squire, Vicki. "Politics through a Web: Citizenship and Community Unbound." *Environment and Planning D: Society and Space*, 30 (2012): 551–67.

Sterne, Jonathan. *The Audible Past: Cultural Origins of Sound Reproduction*. Durham: Duke University Press, 2003.

Stoever, Jennifer Lynn. *The Sonic Color Line: Race and the Cultural Politics of Listening*. New York: New York University Press, 2016.

Stokoe, William C. "Sign Language Structure: An Outline of the Visual Communication Systems of the American Deaf." Originally published in 1960, and reprinted: *Journal of Deaf Studies and Deaf Education*, 10:1 (Winter 2005): 3–37.

tenBroek, Jacobus. "The Right to Live in the World: The Disabled in the Law of Torts." *California Law Review*, 54:2 (1966): 841–919.

Terranova, Tiziana. *Network Culture: Politics for the Information Age*. London: Pluto Press, 2004.

Thompson, Emily. *The Soundscape of Modernity: Architectural Acoustics and the Culture of Listening in America, 1900–1933*. Cambridge, MA: The MIT Press, 2002.

Tiqqun. *Introduction to Civil War*. Los Angeles: Semiotext(e), 2010.

Truax, Barry. *Acoustic Communication*. Norwood, NJ: Ablex Publishing Corporation, 1994.

Tsing, Anna Lowenhaupt. *The Mushroom at the End of the World: On the Possibility of Life in Capitalist Ruins*. Princeton: Princeton University Press, 2015.

Vanbesien, Robin. *Solidarity Poiesis: I Will Come and Steal You*. Berlin: b_books, 2017.

Warner, Michael. *Publics and Counterpublics*. Cambridge, MA: The MIT Press/ Zone Books, 2005.

Weizman, Eyal. *Forensic Architecture: Violence at the Threshold of Detectability*. New York: Zone Books, 2017.

Williamson, Aaron. "Performance, Disability, Humour and Being in Public: Aaron Williamson in conversation with Marquard Smith." In *Aaron Williamson: Performance, Video, Collaboration*. London: Live Art Development Agency, Kiosk, 2008.

Wishart, Trevor. *On Sonic Art*. London: Routledge, 1996.

Wittig, Monique. *The Lesbian Body*. Boston: Beacon Press, 1986.

Wittig, Monique. *The Straight Mind and Other Essays*. Boston: Beacon Press, 1992.

Wright, David. *Deafness: A Personal Account*. London: Faber and Faber, 1991.

Zylinska, Joanna. *The End of Man: A Feminist Counterapocalypse*. Minneapolis: University of Minnesota Press, 2018.

Acknowledgments

During the writing of *Acoustic Justice*, I had the opportunity to collaborate with a number of institutions in presenting and working through the concept of acoustic justice. Such opportunities for thinking and making together are extremely helpful on many levels, from challenging ways of understanding to discovering ideas through particular conversations, as well as keeping the work of discourse close to the sensuality of being together. This included a workshop in Madrid in 2019, in collaboration with the organization AMEE, which led to very enriching and joyful discussions and creative experimentation. I extend my warmest appreciation to Marina Hervás, Víctor Aguado Machuca, Ramón del Buey Cañas, Alberto García Aznar, and all the contributing participants. A visit to Santiago de Chile in 2019 also provided a rich and inspiring platform for presenting and discussing acoustic justice and its relation to the context of Chile. In collaboration with local musicians, composers, artists, and academics, we were able to explore this through a performative workshop and seminar, which deepened my sense for what acoustic practices can do. I'm grateful to the participants, and to Soledad García Saavedra, from the Museo de la Solidaridad Salvador Allende, and Fernando Pérez Villalón, from the Alberto Hurtado University, for their support and collaboration. Two additional opportunities were important in allowing me to present and discuss ideas of acoustics and acoustic justice in a collaborative manner. A workshop held at Konstfack in Stockholm, and one held as part of the Anthropocene Campus, Lisbon, allowed for expanding and testing ideas with others. I'm extremely grateful to Ricardo Atienza and Robin McGinley, and Iñigo Sánchez Fuarros, Daniel Paiva, Ivo Louro, and Margarida Mendes—thank you for your incredible hospitality and engagement. Finally, an edition of the Dirty Ear Forum held at Q-02, Brussels, in 2019, was important in drawing out perspectives on the politics of recognition as being about speaking and listening, as well as about the powers of attention and affection. Thanks to Anna Raimondo, Elena Biserna, Caroline Claus, Jérôme Giller, and Lia García for the extremely resonant and thought-provoking collaboration.

I want to also extend much warmth and appreciation to Luis Guerra, Octavio Camargo, Tao G. Vhrovec Sambolec, Diego Rotman, Jeremy Woodruff, Ricarda

Denzer, Berit Fischer, and Annette le Fort—the capacity for doing is always a family affair. I'm immensely grateful for your caring.

An earlier version of Chapter 2 appeared in the anthology *Sound Art*, published by ZKM and the MIT Press (2019). Chapter 3 took shape following a talk given at the Sound and Participation conference organized by Q-02, Brussels, in 2018. In addition, aspects of Chapter 1 were first discussed in the second edition of my book, *Acoustic Territories*, published by Bloomsbury (2019). I'm grateful for these opportunities, and the support offered in enabling me to work out a range of topics that were helpful in formulating the concept of acoustic justice.

Index

www.ingramcontent.com/pod-product-compliance
Ingram Content Group UK Ltd.
Pitfield, Milton Keynes, MK11 3LW, UK
UKHW020215290926
15470UKWH00001B/191

9 781501 368202